SpringerBriefs in Philosophy

For further volumes:
http://www.springer.com/series/10082

Alex Walter

Evolutionary Psychology and the Propositional-attitudes

Two Mechanist Manifestos

 Springer

Alex Walter
Rutgers University
New Brunswick
NJ 08903, USA

ISSN 2211-4548
ISBN 978-94-007-2968-1
DOI 10.1007/978-94-007-2969-8
Springer Dordrecht Heidelberg New York London

e-ISSN 2211-4556
e-ISBN 978-94-007-2969-8

Library of Congress Control Number: 2011945028

Springer is part of Springer Science+Business Media (www.springer.com)

Preface

While I was still an undergraduate, I was privileged to participate in an interdisciplinary graduate seminar convened by Ray Hyman, Richard Littman, and Doug Hintzman of the Psychology Department at the University of Oregon and Cheyney Ryan of the Philosophy Department there. I recall one day that Hintzman brought in a copy of Dan Dennett's *Brainstorms*. He pointed to a diagram of mental representation constructed by Dennett and remarked that it must have been drawn by a philosopher, since a psychologist would construct a diagram with fewer lines and arrows. He was referring specifically to those lines and arrows that were allocated to beliefs and desires.[1] But then, what alternative is there?

The principal concern of this book is to address the question of how best to characterize proximate mechanisms within the context of evolutionary explanations of human behavior and social organization. The most important task in my view is to evaluate the adequacy of belief-desire, or propositional-attitude, psychology as a role model for proximate level explanation, as this model has come to dominate explanation in the relatively new field of evolutionary psychology. I believe the widespread acceptance of belief-desire psychology by evolutionary psychologists is a big mistake and perhaps an evolutionary dead-end for the discipline. I will argue instead that evolutionists need to embrace the mechanistic perspective. This perspective views the mind-brain as a physical system that operates by means of non-sentence like processes. Accordingly, I will attempt to move proximate explanation away from cognition and towards motivation. That move, however, does not entail that cognition will be ignored. The nature of cognition and its role in the explanation of behavior must be addressed.

When one opens the first chapter in E.O. Wilson's tome, *Sociobiology: The New Synthesis*, he or she must be struck by the architectural magnificence of such a cross-disciplinary structure that encompasses many well established independent or semi-independent disciplines. In this plan, sociobiology is shown to be

---

[1] I refer the reader to Hintzman's excellent review of connectionist models, "Human Learning and Memory: Connections and Dissociations." *Annual Review of Psychology, 41*: 109–139, 1990

intimately connected to various sister disciplines that include cellular genetics, population genetics, and various forms of behavioral biopsychology, including physiological psychology and ethology. Wilson provided a Miroesque diagram with three temporal guideposts indicating how he saw the evolution in the relationships between these disciplines to have proceeded from 1950 to 1975 (the time of conception), and then again from 1975 to the year 2000. Over the course of time, the share allocated to population genetics became larger, while cellular genetics appeared to remain approximately constant.

The one salient feature in the diagram is that connections between the growing field of population genetics and those allocated to behavioral ecology became more pronounced. The bridge between ecology and population genetics, on the one hand, and physiological psychology and ethology, on the other, continues to exist, but the territory occupied by the latter two actually shrank. Some critics have argued since then that population genetics has come to imperialize the behavioral sciences, in fact cannibalizing them in the process. Enter evolutionary psychology. (Although there is no allocation in Wilson's diagram for evolutionary psychology, one can easily see that it would fit in with physiological psychology and ethology—as would cognitive neurobiology.) Its principal architects regard the goal of evolutionary psychology to be a research program capable of providing an account of the evolution of adaptive psychological mechanisms.

Evolutionary psychology has proven to possess imperial ambitions of its own, attempting to displace sociobiology as the paradigm of choice. This power grab is motivated by the cannibalization of psychology by population genetics at the hands of sociobiologists. Evolutionary psychologists are attempting to reclaim lost (or stolen) territory. Several commentators, such as Irons (1989) and Blurton Jones (1989), have tried to paint the internecine dispute as a false dichotomy, but evolutionary psychologists tend to defend their turf, arguing that adaptive mechanism is key and that sociobiologists have irredeemably abandoned this element of explanation.

My objective is to build a charter for a mechanist approach to the evolutionary analysis of the psychological mechanisms that underlie behavior. The information processing model that currently holds sway amongst evolutionary psychologists is fundamentally inimical to a materialistic view of life, mind and behavior. I think the appeal of the information processing model is because it appears to provide a fast route to scientific progress. This is understandable when a discipline is comparatively young and surrounded by hostile forces. Of course, one wishes to be a member of what Imre Lakatos described as a progressive research program rather than a member of a degenerating one. And while I even advocate Feyerabend's twin principles of proliferation and tenacity, especially for young scientific disciplines, at some point I think progress can best be made by shoring up one's metaphysical commitments—if only because these come to hinder the development of conceptual coherence in the research tradition. In short, although of possible heuristic value in the early going, I think the cognitive turn in evolutionary psychology is a mistaken commitment that needs to be addressed. I think we need to focus on motivational endowments at the level of neural mechanisms, but the evolutionary history of such mechanisms is an essential part of the story and cannot be neglected. I think it is time we attempt to conceptualize proximate and ultimate forms of causation in terms consistent with philosophical materialism, not in terms consistent with mind-body dualism.

In the first chapter, the commitment of evolutionary psychology to information processing metaphysics, more specifically mind-body dualism, is examined and critiqued. The second chapter is a logical extension of the first. The focus of the second essay is on the key role that motivational endowments play in the explanation of behavior. Reinforcement theory plays a major role in the arguments presented in the both these essays. Reinforcement theory provides an explanation of non-cognitive aspects of motivational endowments. A mechanist account that includes reward event theory challenges the basic tenets of folk psychology. The third chapter attempts to find a coherent path through the modularity issue, examining the domain-specificity arguments advanced by evolutionary psychologists, as well as those provided by theorists offering a cognitive development approach.

Acknowledgments

I would like to thank the following persons for encouragement and assistance over the course of this project. I'd first like to thank Robin Fox, Dieter Steklis, A.P. Vayda, & Lionel Tiger, who served as mentors in graduate school at Rutgers. Professor Tiger made particularly useful comments on the present project. Brad Walters, Mt. Allison University, New Brunswick, Canada, provided helpful suggestions on repeated occasions during the development of this work. I also wish to thank Martie Haselton at UCLA for her incisive comments on an early draft. The work has benefitted from discussions with Anthony Biglan, Paul Churchland, and Stephen Stich. I'd especially like to thank my father-in-law, Dr. Taya Alami, Emeritus Professor of neuroanatomy, University Muhammed V, Rabat, Morocco, who crafted the drawing of the limbic system that appear in the second essay. Portions of Chap. 3 are drawn from my article, "Stone Age Detective's Paradise," Reviews in Anthropology 31(1), 2002, 29–61. Permission granted by The Taylor Francis Group.

Contents

1   **Evolutionary Psychology and the Propositional-Attitudes: Why the Cognitive Turn is a Wrong Turn** 1
    Varieties of Propositional Experience. 1
    The Intentional Bogeyman 2
    Sociobiology and the Propositional-Attitudes 7
    The Cognitive Turn in Evolutionary Psychology 9
    The 'Intentional Systems' Model of Mental Evolution 10
    The Information Processing Model 12
    Propositional-Attitudes and Mind/Body Dualism 17
    Intrinsic Intentionality and Folk Psychology 22
    Folk Psychology: The Best or Worst of Intensions?. 26
    Conclusion: The Mechanist Stance 27

2   **Why I am not an Evolutionary Psychologist: On the Imperative Nature of Motivational Endowments.** 29
    Motivational Endowments 30
    Proximate Mechanisms: The Central Role of Emotion, Drive and Reward 31
    Are Emotions Proximate Mechanisms? 34
    The Ghost in the Reward Mechanism 37
    Why Proximate Mechanisms Matter 42
    Folk Psychology Redux 43
    The 'Psychologic Gambit' Versus the Mechanist Stance 44
    Evo-Devo Versus Evolutionary Psychology?. 45
    The Physiologic Gambit in Evolutionary Psychology............ 47
    Conclusion: The Psychologic Gambit in Decline or Reductionism at Last! 49

3 Postscript: The Virtues of Weak Modularity................ 51
   Portrait of the Mind–Brain as an Evolved Epigenetic System....... 54
   Hominid Evolution and Cognitive Fluidity.................. 56
   Weak Versus Massive Modularity....................... 60
   Modularity and the Theory of Multiple Intelligence............ 63
   Explaining the Evolution of the Pleistocene Mind–Brain.......... 65

Epilogue... 69

References.. 73

Index... 83

Chapter 1
Evolutionary Psychology and the Propositional-Attitudes: Why the Cognitive Turn is a Wrong Turn

What is a propositional-attitude and why does it appear to be important to psychology? A diverse array of philosophers and psychologists—although they agree on little else—assure us that explanation in psychology can not exist without a strong commitment to such attitudes. In a nutshell, 'propositional-attitude' refers to a model of mental life that is presented as though the structure of thought is that of sentences. Jerry Fodor, foremost among the various advocates of this commitment, has stated the case for the centrality of propositional-attitudes, first in his seminal *Language of Thought* (Fodor 1975), and then again in his reprisal, *LOT 2* (Fodor 2008). The basic form of the model was sketched in *Language of Thought* (Fodor 1975) as follows: An agent finds himself in situation S. The agent believes that a certain range of behavioral options (b_1 b_2... b_n) are available in situation S. The probable outcomes that follow from performing each b in S are calculated (including relevant *cetus paribus* clauses). A preference ranking is matched against the agent's preference standards. The agent makes a decision that optimizes the desired outcome given the preference standards and the probabilities assigned to the different outcomes (Fodor 1975, pp. 28–29). Although Fodor attempts to convince the reader that *there must be a* 'language of thought', I would like you to consider the possibility that there is no such thing as a 'language of thought' because the premises of the argument are deeply mistaken.

Varieties of Propositional Experience

Propositional-attitudes come in two basic forms. The first is Folk Psychology which is generally construed on the model of conscious *intentions* (Searle 1980a, 1980b, 1984), *reasons* (Davidson 1963, 1980; Toulmin 1970), or rational practical reasoning (Driscoll and Stich 2008). The second set consists of Information Processing Models (Chomsky 1965, 1972, 1980; Fodor 1975, 2008) Does either

A. Walter, *Evolutionary Psychology and the Propositional-attitudes*, SpringerBriefs in Philosophy, DOI: 10.1007/978-94-007-2969-8_1, © The Author(s) 2012

model provide a suitable candidate for an explanatory model in evolutionary psychology? Does either provide an account of the *proximate mechanisms* that produce behavior? My Argument is that both do not. Both fail, however, for different reasons. Folk Psychological (FP) models fail because motivational endowments include mechanisms that are *subdoxastic* (Stich 1978), and because such mechanisms must be related to their evolutionary and proximate stimulus conditions. Information processing models (IPM) fail because they are dualistic since they posit a level of mental events that violate philosophical materialism (Feyerabend 1963a, b). That is, IP models are either explicitly or implicitly committed to mind–body dualism as argued by Uttal (2004, 2005). Finally, IP models also fail because representations in the brain are not sentential or propositional in nature (Churcland 1986; 2007). Despite their differences, however, both forms of propositional-attitude have at their base a commitment to intentionality.

The Intentional Bogeyman

What is *intentionality* and where does it fit into the explanation of behavior? Dennett (1983, 1996) advanced the thesis that there were three 'stances' one could strike in theX explanation of behavior: the physical, the design, and the intentional. The physical stance is materialist—it provides explanations in terms of physics, chemistry, and other biological structured wholes that supervene on physics and chemistry, such as genes or neurons. The physical stance specifies what the sociobiologist or evolutionary psychologist calls the proximate mechanism. The design stance, on the other hand, specifies the adaptive history that explains the function that the physical thing is 'designed' to perform. The design stance specifies the *adaptation*. The design stance would thus specify a creator/designer or a process such as natural selection. Hence, on the physical stance we would specify, for example, the material mechanisms and structure of an electric can opener, while on the design stance we would specify that its purpose was to open cans. If we are dealing with a biological system, the 'designer' is natural selection and the proximate mechanisms might include a neural system that produces a given behavior or the genes that catalyzed the development of that particular structure. Although Dennett asserts that the physical stance trumps the design stance, it should be clear to sociobiologists or evolutionary psychologists that the design stance is not inferior to the physical stance. Both have different and complementary roles to play in the explanation of behavior. Since evolutionary psychologists (and sociobiologists) believe that proximate and ultimate categories of explanation are mutually exclusive and exhaustive, have they overlooked an entire domain of explanation called the 'intentional'?

I have not yet stated what an *intention* is. Dennett informs us that the philosophical meaning of the term is to be distinguished from the commonsense meaning, which refers simply to the deliberate nature of an action. For instance, you intended to open the door and not the window. (We will discuss the nature and place of commonsense intentionality below when I address John Searle's notion of *intrinsic* intentionality.) The philosophical notion of intentionality has several parts, all of which are a bit slippery in that one slides into the other, forming what I call the 'slippery slope of intentionality'. The foremost feature of the intentional, according to Dennett (and this is also asserted by Jerry Fodor), is its *aboutness*. An intention is a mental state that is 'about' something. This is where the slippery slope begins and it is a steep one because we are all of a sudden introduced to the apparent necessity of the propositional-attitudes of *belief* and *desire*. For instance, if I desire a roast beef sandwich and I believe there is roast beef in my refrigerator, I may go into the kitchen with the intent of making a roast beef sandwich. In fact, belief + desire = intention. The belief about the existence of the roast beef and the desire for it constitute an intention to make and eat a roast beef sandwich. Each of these propositional-attitudes is manifest in the sentences in which they are expressed. What more could we want in an explanation than that?

The puzzling thing about this type of intentionality is that animals that do not exhibit language abilities also can be said to harbor intentional states, although Dennett refers to the attribution of intentionality to non-linguistic creature as 'as if'. Even though hornets and butterflies do not have verbal capabilities, they can be said to behave 'as if' they had beliefs and desires. This type of thinking is prevalent in sociobiological thinking even today. For example, Hales (2009) quotes Joyce (2006) as characterizing kin selection in just such an 'as if' form of intentional state: "provide-help-to-those-conspecifics-with-whom-you-interact-frequently" (Joyce quoted in Hales 2009, p. 437). While this sort of 'as if' rule-following behavior may strike some as plausible in the case of sentential being such as humans, it becomes somewhat more contentious when applied to non-sentential creatures such as crocodiles and birds. Hales shows how absurd it would be to characterize a situation of reciprocal altruism between a crocodile and a ziczac bird: "Ziczacs and I [croc] have been engaged in a cooperative enterprise for a while now, and I have every reason to believe that we will be business associates well into the future". "Since our relationship has been so mutually advantageous so far, I choose not to eat a ziczac today" (Hales 2009, p. 440). Dennett would assert that this 'rule' is just a short-hand way of speaking about this relationship 'as if' these creatures had intentionality even though we know some causal-mechanical account is required instead. Dennett readily admits that the design and physical stances trump the intentional stance, but is the intentional stance a coherent way to stand? As Patricia Churchland (1983) points out in her critique of Dennett's 'as if' intentionality, the frog at the bottom of Dennett's cup is the commitment to propositional-attitudes which are necessarily based on some form of language of thought model.

Ruth Millikan is a philosopher very close in thinking to Dan Dennett. (In fact, on the back of her book of essays, *White Queen Psychology and other Essays for Alice* (Millikan 1993), Dennett states: "[T]his is how philosophy of psychology should be done!" Millikan is a pioneer in the field of biosemantics, and making the intentional stance legitimate is one of her primary objectives dating from her rigorously formulated tome, *Language, Thought and Other Biological Categories*

(Millikan 1984). Less rigorous but perhaps more revealing are the arguments she makes on behalf of Dennett's 'as if' intentionality in *The White Queen*. To illustrate, I will make use of just one example: the well known honey bee waggle dance discovered and described by the ethologist Karl von Frisch. The mechanics of the dance are quite complex involving a number of stimulus–response variables. To simplify, the direction the bee moves in relation to the hive indicates direction. If it moves vertically upwards, the direction to the source is directly towards the Sun. The duration of the waggle part of the dance signifies the distance (Von Frisch 1967, pp. 61–63).. Bees who witness this dance are able, on the basis of the dance, to locate the honey indicated in the dance steps. Certainly, this bee behavior has an adaptive explanation, and possibly someday the precise neural mechanisms will be accounted for, but where does the intentional stance fit in? The intentionality consists first in the fact that the dance is *about* the location of the honey. But since intentions = beliefs + desires, the explanatory language of propositional-attitudes has to be applied, even if it is clear that honey bees are incapable of formulating the sentences in which beliefs and desires must be expressed. Do the bees who witness the dance *believe* they know where the honey is? Do they *desire* to obtain the honey? What do they have to do to find it? Millikan makes it clear that she does not think that bees have beliefs or desires in the way that human beings have beliefs and desires because of the non-sentential nature of their inner representations. She says that bee beliefs and desires are like a toad's beliefs or desires. For example, the toad desires flies and believes (mistakenly) that the bbs the experimenter is trying to roll past him are flies and so flicks out its fly getting mechanism (i.e. tongue) to obtain the flies (Millikan 1993, p. 78). Now, instead of beliefs and desires, Millikan substitutes *indicatives* and *imperatives*. The indicative signal tells the bee where the honey is, while the imperative supplies the motivation to retrieve the honey. She conceptualizes indicatives and imperatives as some type of internal map (although she does not indicate what type of map that might be). Nevertheless, I think indicatives and imperatives are a better way to characterize the functional relationship between the inner representations of the behaving organism and the environmental stimulus because these potentially avoid the commitment to a theory of representation based on propositional-attitudes. This is true in part because the imperative half of the equation is inherently non-cognitive in nature. The task is to show that the indicative component is non-intentional as well.

Sterelny (2003) has argued on behalf of the intentional paradigm that intentionality resides in the decoupling of indicatives and imperatives (pp. 26–29). Sterelny grants that in respect to simple signal systems, where the response is dictated by the reception of the stimuli, no intentionality is required. However, in tracking systems that employ multiple reception systems, the response is contingent upon rule-governed evaluaton. Using the example of the cognitive systems that bees use to locate food in response to the location 'dance' (vide Gould and Gould 1988), he argues that a set of nested and contingent hierarchical evaluations can be made depending on the circumstances. Honeybees' first preference is to use landmarks. If these are not available, they will use their evolved solar compass, and if it is too dark to use the latter they will use a mechanism that responds to the polarization of available light (Sterelny 2003, pp 23–24). Sterelny refers to the response biases sensitive to this contingent set of circumstances as "rules of thumb" (p. 24). The claim is that bees are following "rules of thumb" when they decide to use one set of cues rather than another, and Sterelny characterizes these rules as "belief-like" representations (p. 31). That different location mechanisms are activated under specific stimulus conditions is no doubt a product of an adaptive history, but the appeal to intentionality seems farfetched. We need an explanation that characterizes the proximate mechanisms in terms of Dennett's physical stance. Bee honey location "rules of thumb" appears to be an example of Dennett's 'as if' intentionality.

Although Sterelny rejects Churchlandian models to cover the domain of indicatives (2003, p. 96), I will argue below that some type of neurobiological theory of representation, such as that advanced by Paul and Patricia Churchland (1998; Churchland 1986; 2007), or, alternatively, Grossberg (1988),[1] provides a better account of how the brain 'maps' indicatives than does propositional-attitude psychology.[2] I also intend to argue that non-propositional models of representation are even the case for sentient, language wielding beings, namely, *Homo sapiens*. Moreover, I intend to show that intentional explanation, in the philosophical sense of the term, as practiced by its adherents in both philosophical psychology and evolutionary psychology provides a confused and unnecessary half-way house between proximate mechanistic explanation and ultimate (i.e. adaptive) explanation.

This section is titled 'The Intentional Bogeyman' and while the intentional stance might be inadequate or misguided in its quest to supplement proximate and ultimate explanation, the 'bogeyman' designation might seem unduly dismissive. Perhaps, if divested of its propositional-attitude commitments, it might have something to contribute to the explanation of behavior. My position is that it can not be divested of such commitments. However, there is another issue that I think provides a shortcut to ridding ourselves of intentionality. That shortcut concerns the metaphysical nature of the intentionality concept. In *Kinds of Minds* (1996), Dennett reminds us that we inherited the concept of intentionality from the philosophical perspective of medieval monks such as St. Thomas Aquinas.[3]

---

[1] I refer the reader to Hintzman's excellent review of connectionist models, "Human Learning and Memory: Connections and Dissociations" *Annual Review of Psychology*, 41: 109–139, 1990. Leading contenders for the model of choice in representational simulation include Parallel Distributed Processing (PDP) (e.g. the Churchlands) and Adaptive Resonance Theory (ART) (e.g. Grossberg). I will not make a commitment to either connectionist school here; the objective of this essay is to argue on behalf of biologically realistic theories of cognition against propositional-attitude based models.

[2] In the second part of this essay, I will argue that imperatives are best modeled on non-propositional theories of motivational endowment proposed by researchers such as LeDoux (1996), Pfaff (1999), and Panksepp (1998).

[3] See St. Thomas Aquinas' *Summa Theologica*, Part I.

Dennett passes through the point casually, but the source of the concept is crucial to understanding its intent (in the commonsense sense of the term.) Medieval Christian philosopher monks were neo-Platonists who believed that everything in the world corresponded to an idea in the mind of God, and that the 'idea' was the reality or essence as distinct from the thing that the idea is about. The defining 'aboutness' of the intentional object is not *about* the physical object in the world, it is about the *idea* of the thing. The important thing to remember is that the 'idea' referenced here is not the idea in the mind of the thinker, it is the objective Platonic essence. Medieval monks thought, somewhat contrary to Plato, that these intentional objects were ideas in the mind of the Christian God. Human 'ideas' are conceived as copies of the universal objective ideas, whether they be conceived as Plato's essences or those of St. Thomas Aquinas. So if you want an easy route out of the intentional realm, there it is.[4]

Platonism continued into the twentieth century, where logical positivists such as Frege held that thoughts are not identical to the declarative sentences by which they are expressed. Rather, 'Thoughts' refer to the *pre-existing objective content* that is expressed in language (Wettstein 2004, pp. 1–66). When some philosophers opted for psychology over metaphysics, as exemplified in Fodor's 'Language of Thought' model, 'thoughts' continued to be conceived as timeless Platonic essences, but they were also simultaneously conceived as psychological events in the minds of thinkers (Fodor 1975, 2008). Similarly, the theory that underlies Saussure's concept of the linguistic sign is that the *signifier* (the word) does not point to the object that it refers to; rather that which is *signified* is the idea or image of the object (i.e. the mental representation).[5] Hence, 'intentionality' came to be applied to the psychological representations of human beings and even honeybees, and philosophers such as Dennett also apply it to humanly designed physical systems such as alarm clocks, guided missiles, or chess playing computer programs (Dennett 1996, p. 30; see also Millikan 1993). An alternative path exists in philosophy that opposes this neo-Platonic view; it has a impressive pedigree that is well represented by individuals such as Ryle (1949), Wittgenstein (1953), Place (1956), Smart (1959), Quine (1960) and, more recently, Churchland and Churchland (1998). For a rare addition from the continent there is Derrida (1973).[6] The members of this group all believe that no ghostly intermediary need be postulated between the object and the word that refers to it.

---

[4] I made a similar anti-metaphysical argument against the naturalistic fallacy in my interpretation of Hume's Law. See my essay, 'The Anti-naturalistic Fallacy: Evolutionary Moral Psychology and the Insistence of Brute Facts, *Evolutionary Psychology*, 6: 32–47, 2006.

[5] See de Saussure's *Course in General Linguistics* (1916\1998) or interpretations of his theory of the sign as offered by Culler (1976) as well as evolutionary theorists such as Deacon (1997, p. 69).

[6] Although some readers might find Derrida to be an odd addition to this group, I refer those who are puzzled to Henry Staten's *Derrida and Wittgenstein* (1986). Derrida relates the relentless search for timeless essences in Husserl, as well as in others, to the authors' fear of death.

The fact that intentional objects show great variation in their *intensions* (i.e. meanings) both within and across individuals does not appear to strike Dennett as a problem for the theory of intentional systems. Dennett says it would be nice if there were fixed, neutral, eternal criteria for fixing the meanings of intentional objects, but he admits this goal is unobtainable simply because of that variation (Dennett 1996, p. 48). That is the point that Wittgenstein repeatedly strove to drive home in his *Philosophical Investigations,* and it is the point that Quine repeatedly comes back to from *Word and Object* all the way through his last collection of essays, *Confessions of a Confirmed Extensionalist* (Quine 2008). According to Quine, what we need to divest ourselves of is the "idea idea" (Quine 1987, p. 89). Against this anti-metaphyiscal stance, Plato, St. Thomas Aquinas, and Jerry Fodor would insist that if you are talking about representations that vary, then you are not talking about intentional objects because such objects must be uniform since they are Platonic essences. I think the clear implication here is that whatever we are talking about when we talk about actual representations in living brains, we must not be talking about intentional objects.[7]

Sociobiology and the Propositional-Attitudes

Let us try to understand why evolutionary psychologists, as well as sociobiologists before them, bought into the propositional-attitude model of representation in the first place. From the outset, sociobiologists took care to draw a fundamental distinction between *ultimate* and *proximate* levels of causation in explaining behavior, the former referring to the selection pressures that produced traits (including proximate mechanisms), while the latter referred to the internal neural mechanisms themselves (Wilson 1975; Barash 1977). Nevertheless, sociobiological explanations tended to "leap frog" over the brain and substitute population genetic explanations in the frame where brain mechanisms should be specified (Churchland 1983). Hence, sociobiological explanations tended to focus on fitness consequences to the exclusion of proximate mediating mechanisms, and this exclusion seemed to result in the attribution of motivational agency to genes themselves (Dawkins 1976). As a result, genes being replicated from one generation to the next were characterized as "selfish", apparently seeking to maximize their representation in succeeding generations. Dawkins tried to make it clear that

---

[7] I discuss the problem of essentialism in respect to the meme concept in my paper, 'The Trouble with Memes: Deconstructing Dawkins' Monster', *Social Science Information*, pp. 691–709, 2007. Dennett (1995) endorses the meme concept without realizing that in doing so he has violated the strictures of Darwin's 'dangerous idea', i.e. that species are not pre-established essences, and neither are memes. It is perhaps even more astonishing that Dawkins himself, the author of The God Delusion (2006) fails to see the elements of Platonic idealism that underlies the meme concept. (One might say that Dawkins got rid of the thinker, i.e. God, but kept the thoughts, i.e. memes.).

'selfishness' was not a motivational attribute of genes but a metaphor for under-standing the effects or outcomes of natural selection or other factors that altered gene frequencies in populations. Dawkins tried to explain how, for example, genes could be selected that made individuals make discriminations in favor of other individuals which are likely to contain copies of the same genes. This is where the appeal to propositional-attitudes emerged in the explanation. In *The Selfish Gene* (1976), Dawkins proposes that 'as if' rules can be understood as if they were 'higher level social strategies' that are programmed into the nervous system such that lower level behavior responses can be left open. Dawkins employs the computer chess playing program as an analogy. The computer is programmed with an instruction such as "keep and/or establish an open file". This can be done through a varied combination of moves, and the computer is programmed to choose alternatives that best correspond to the situation represented on the board. Similarly, organisms are programmed with higher level strategies such as "help your siblings twice as much as your first cousins" but have conditional flexibility in determining the means given the circumstances on the ground. Dawkins suggested that what evolved to facilitate the process was "a rough rule of thumb such as share food with anything that moves in the nest in which you are sitting" (Dawkins 1979, p. 187). Sterelny (2003) follows Dawkins' use of 'rules of thumb' to characterize intentional mental states. According to Sterelny, creatures that have decoupled representations and have achieved response breadth are not tied to immediate stimulus–response relationships (Sterelny 2003, p. 76). They can use what information they have in their cognitive storage systems to make decisions. They act on rules rather than respond to cues.

This way of thinking had a predecessor. Prior to the emergence of sociobiology, Tiger and Fox (1971) (cf. Tiger 1994) proposed a model where functional goals could be achieved by diverse behavioral means that was based on Chomsky's generative grammar model (Chomsky 1965, 1980) which they referred to as 'biogrammar'. In the case of Tiger and Fox, Chomsky's model seemed to provide a solution to how group selection worked. If the survival of the group could be accomplished by various means (*p, q,* or r), then these various means were merely instrumental in achieving their functional end (i.e. group survival), and it did not matter which means was employed. The evolutionary goal was accomplished roughly in a way similar to how different sentence constructions could be used to say the 'same' thing. Biogrammar was thus construed as the proximate mechanism by which group selection worked. The selection of Chomsky's grammar theory was based on the latter's popularity and the ease with which it could be adapted to the explanatory purpose—that is, if you were not too particular about its actual relevance to non-linguistic behavior. The appeal to propositional-attitudes in the case of Tiger and Fox was therefore largely serendipitous. However, since Dawkins (1976), the software analogy is construed as the paradigm to model behavior as a function of proposi-tional- attitudes. This was, to say the least, a very soft approach to the proximate mechanism issue. The 'as if' intentional explanation utilized propositional-attitudes in the form of beliefs and desires "as if" these were the *bona fide* proximate mechanisms that are required by the evolutionary explanation.

The Cognitive Turn in Evolutionary Psychology

Evolutionary psychology came to supplant sociobiology as the dominant paradigm in the evolutionary explanation of human behavior because its adherents claimed to provide a more adequate account of adaptive proximate mechanisms. These innovations have been regarded as constituting a major paradigm shift in the evolutionary study of human behavior. The most fundamental criticism has to do with the issue of the number and specificity of cognitive adaptive mechanisms. Sociobiologists were held to believe that the mind-brain consisted of one domain-general fitness maximizing calculator, while the new breed of evolutionary psychologists posited a large number of domain-specific adaptive modules (Symons 1992; Buss 1991, 1994; Cosmides and Tooby 1992, 1994; Tooby and Cosmides 1989, 1990; Pinker 1997, Miller and Todd 1998).

Buss (1991) and Tooby and Cosmides (1990) expressed doubt that a domain-general inclusive fitness-maximizer exists because only correlates of fitness can be tracked during a given lifetime. Hence, human psychology is geared towards obtaining rewards that historically (in ancestral environments) conferred fitness. As a result, specific adaptive tasks needed to be met and this resulted in domain-specific psychological mechanisms rather than a general fitness calculator (Buss 1991, p. 463). Buss defines a psychological mechanism as an internal process that is keyed to a specific problem related to survival or reproduction. Hence, it is sensitive to specific types of input from which it extracts the relevant data, processes that data, resulting in a decision-rule that produces or regulates behavior which is geared towards obtaining an adaptively relevant outcome (Buss 1991, p. 464). What one should note on this model is that it is also based on a computational model of information processing. It is just that the tasks are specific rather than general.[8]

For the evolutionary psychologist, it appears that proximate psychological mechanisms come in an assortment of forms: dispositions, decision-rules, structures, processes, etc. (Buss 1991, p. 461). Some may involve propositional-attitudes while others may not. The ones that do, require the manipulation of various kinds of information due to the existence of situational complexity where a number of factors have to be considered and which thus require a decision rather than just an unconditioned response (i.e. reflex) or a conditioned (i.e. learned) response. It thus seems that proximate mechanisms form a highly problematic hodge-podge of very dissimilar types of entities.

While evolutionary psychologists see themselves as improving on the original aspirations of sociobiology, hostile critics of sociobiology, such as the human ecologist, Vayda (1995, 2008), also refer to the role of propositional-attitudes as constituting fundamental psychological mechanisms. These include: "the intentions, purposes, knowledge, and beliefs on the basis of which particular actions are undertaken... and... the psychological and neurophysiological mechanisms that

---

[8] For a further discussion of the modularity thesis see Chap. 3.

trigger particular behavior in particular contexts" (1995, pp. 273–274). While Vayda categorically asserts that sociobiologists ubiquitously ignore such proximate mechanisms, he admits that sometimes evolutionary psychologists provide them. Hence, friend and foe alike seem to agree that proximate mechanisms are a mixed bag that contains both subdoxa and propositional-attitude.

The evolutionary population biologist, D.S. Wilson, collaborated with the philosopher of science, Elliott Sober, in an attempt to formulate a proper account of how proximate mechanisms should be conceived. Sober and Wilson (1998) argued that belief-desire psychology is indispensible in accounting for proximate psychological mechanisms. They both agree that "an evolutionary perspective on human behavior requires us to regard the human mind as a proximate mechanism for causing organisms to produce adaptive behaviors" (Sober and Wilson 1998, p. 200). The form of the proximate mechanism for which they advocate is the intentional. They argue, for example, that the proximate mechanism that produces bodily injury avoidance behavior is the *desire* to avoid pain (Sober and Wilson 1998, p. 201). Elsewhere, Sober has suggested that the proximate mechanism for avoiding incest and its deleterious consequences is the *idea* that incest is bad and that this *idea* performs the same function in the human case that a neurological set-up functions in the case of a non-human animal (Sober 1985, p. 188). In other words, in species that do not have language, the proximate mechanism is a subdoxastic physiochemical process, while in species that do have language, the mechanism is transferred to the resulting belief or desire.

Against this view, I wish to demonstrate that propositional-attitudes are not *bona fide* proximate mechanisms, but that they should instead be regarded as behavior. Proximate psychological mechanisms consist of physiochemical, non-sentential, subdoxastic processes that generate behavior, and such mechanisms are the same for language bearing species, as well as for non-language bearing species. Hence, a fundamental explanatory distinction needs to be drawn between *behavior*, which might include verbal behavior that expresses propositional-attitudes, and subdoxastic motivational mechanisms, which include reward systems in the nervous system that are non-propositional in nature.

The 'Intentional Systems' Model of Mental Evolution

In *Kinds of Minds* (1996), Dan Dennett provides a progressive contextualization of the kinds of minds that he thinks evolution has produced. These appear in evolutionary sequence as: Darwinian, Skinnerian, Popperian, and Gregorian.[9] The primary emergent characteristic in this series, according to Dennett, is that each

---

[9] Although Dennett invents the term, *Gregorian*, in honor of the neuroscientist, Richard Gregory, I like to think that Dennett invented the term while listening to Gregorian chants and drinking wine out of a medieval goblet.

successor can do things that its predecessor can not. Each successive accomplishment in the chain results in greater complexity and self-determination. Simple Darwinian critters, such as bacteria and barnacles, are incapable of learning anything from experience that could modify their behavior. Skinnerian critters are those that are capable of trial and error learning. This would include most species, including *Homo sapiens*, but while most creatures that can learn are limited to trial and error learning, and are hence strictly Skinnerian critters, Dennett goes on to argue that *Homo sapiens*, and possibly certain other species, exceeds those limits because they can learn from the experience of others without having to figure it out for themselves based on their own encounters with the stimuli they confront. These latter are Popperian critters. Finally, critters that can see how to use tools to achieve their ends ascend to the top level of the hierarchy. (These are the Gregorian critters.) Although, a small number of species can be said to use tools in at least a rudimentary way, such as chimpanzees, the most advanced tool is language and this is restricted to *Homo sapiens* and, with caveats to the rudimentary nature of their nascent abilities, our ape cousins. Clearly we are far down the path that permits us to escape from blind mechanism and achieves consciousness and, with it, the wonder of intentionality. With language one can form propositional-attitudes, that is, beliefs and desires and, as we learned above, belief + desire = intention. Dennett's claim is that we Gregorian critters have outdistanced mere Pavlovian and Skinnerian conditioned critters.

Dennett deserves credit for recognizing that operant conditioning is still at work even in critters as intentionally marvelous as our species, but it is not clear that he has taken the Skinnerian ball as far as it can go. Skinner made it clear in *Science and Human Behavior* (1953) that human beings are capable of learning from the reinforcement histories of others without having to experience contingencies of reinforcement directly themselves. He argues that this is what 'culture' is. Similarly, following a 'rule' is largely a matter of taking what is learned from the experiences of others and formulating it as a guide to action. Such shortcuts enable us to obtain positive rewards and avoid punitive consequences. Hence, Popperian critters turn out to be a sub-species of Skinnerian critter.

Let us cut to the chase. How about language bearing Gregorian critters? Certainly Skinner thought he gave an account of how language is a tool in the operant control of behavior. He considered *Verbal Behavior* (1957) to be his masterpiece. It was the one work of his that applied specifically to *Homo sapiens*. He attempted there to show how operant conditioning could account for important features of linguistic behavior, including grammar. On this model we would have to reverse the relationship of Dennett's hierarchy of critter types. All Gregorian critters belong to the subset of Popperian critters and all Popperian critters belong to the subset of Skinnerian critter. In turn, since all critters are the product of Darwinian evolution, all Skinnerian critters belong to the subset of Darwinian critter. Of course, the means and degree of flexibility of response differ from species to species, but the question remains as to whether the Gregorian critter has escaped causality through language. This is frequently held to be one of the primary accomplishments of language versus other systems of behavior. Language promotes response breadth and facilitates the decoupling of representations that is required by Sterelny (2003) for *bona fide* intentionality. A *decision* maker simply does not respond to stimuli; the decision maker evaluates the various contingencies and decides which stimuli are relevant and then chooses the appropriate means to achieve the actor's chosen goal. Sterelny therefore argues, similarly to Dennett, that the attainment of decoupled representation and response breadth turns the table on subdoxastic motivational endowments. It is 'as if' *free will* had evolved.[10] How could a behaviorist escape from such a secure half-Nelson hold? One way would be to point out that we are still Skinnerian and Darwinian critters and that decoupled representation and response breadth, although more complex than reflexes, still operate in the realm of mechanism. Even though the deterministic pathways become more complex, perhaps intractable, there is no principled recourse to free-will.

Now, of course, Skinner's account of language met with stiff opposition, and this stiff opposition came in the form of Noam Chomsky's 'transformational generative grammar' paradigm (Chomsky 1959, 1965, 1972, 1980). Although Chomsky's 1957 critique of Skinner was primarily a mistaken diatribe against Pavlovian conditioning in the domain of language (MacCorquodale 1970), the resulting conceptual revolution turned attention away from the extensional realm of behavior and towards the internal realm of cognition. Cognition was thereafter conceiving as an internal map of rules and representations, in other words: propositional-attitudes. That is why Jerry Fodor claims that there *must* be a 'language-of-thought'; there must be because language is the nature of thought. There are, however, good reasons for believing that language is not a good theory with which to model the nature of thought, and that these reasons apply to language itself. What kind of an objection could be given to such an obvious truth? According to LOT theorists, the language we speak must be a direct expression of its internal structure and dynamics. But is it?

The Information Processing Model

The Information Processing Model (IPM) is explicitly predicated on sentential attitudes as the 'language of thought'. It partakes of the same model of the logic of propositional-attitudes presented in the quotation drawn from Fodor under section "The 'Intentional Systems' Model of Mental Evolution". The model is based on sentences and logical relations between sentences (see Churchland 1980a, p. 150). However, instead of being applied to a conscious reasoning sequence, the model assumes that these sentential attitudes occur non-consciously at the level of pre-processing. Hence, the overt spoken sentence is the outward manifestation of an

---

[10] Dennett attempts to make this argument in his *Freedom Evolves* (Dennett 2003). The view of evolved freewill that Dennett advances is that known as 'compatibilism', the view that freewill is compatible with determinism. Dennett calls it a freewill that is worth having. Behaviorists and other determinists do not think that sort of freewill is freewill.

internal sentential computational process. On extreme versions of this thesis, in the case of human beings, even non-propositional cognitive representations such as those that occur in images. In the case of non-language bearing species, alternatively, sentences are used to model their intelligent behavior. This universal cognitive code has come to be known as 'mentalese'. Mentalese is a substitute term for Chomsky's theory of a universal grammar (Chomsky 1965, 1980); it is the ur-language that underlies and precedes the acquisition of an actually existing human language. For those committed to strong IPM, this model underlies all thought of all species. If the model is true, then it might make some kind of sense to model social relationships between Ziczac birds and crocodiles on mentalese. If not, then some rethinking needs to be done about Dawkins style fitness oriented cognitive programs.

The issue here is whether or not explicit sentential attitudes get more credit than they deserve. IPM makes sentential brain products dominant over non-sentential processes. Patricia Churchland argues that language is just one of the types of activity the brain engages in, and it makes little sense to presuppose that all activities can be modeled on linguistic processes. Churchland summarizes the stakes as follows:

> The query is this: are inner processes best thought of as analogous to action performed with speech albeit silently and unconsciously; and consequently, should the processing be taken to consist pretty much of complex sequences of sentential attitudes (believes that p, calculates the probability of e in m/n, wants that p, and so on mediated by such operations as inference? Or alternatively, is it possible that the best theory will be one which hobbles the model of overt speech, relegates sentential attitudes of very minor role in the theory information processing, and seeks in their stead non-sentential states and processes to account for how the brain represents the world and guides behavior. (P.S. Churchland 1980a, 149.)

As Patricia Churchland reveals, the strong form of the sententialist thesis comes in several varieties. The first variety consists of models that do not include a non-sentential processing component. All processing is some type of pre-processing (which may be sentential or not) plus the sentential component. Here pre-processing component appears to be some kind of data set that is acted upon by sentential attitudes. The difference between the sentential attitudes is whether there is a commitment to mentalese or whether the sentential component consists solely of a real existing language. Chomsky himself might be included in the subset of sententialists that divide the IPM component democratically between mentalese and real acquired languages.

The second group hypothesized by Patricia Churchland reduces the extent of the cognitive geography that is devoted to sentential attitudes and includes a larger geographic proportion to non-sentential attitudes. She argues that there are three basic reasons for privileging non-sentential attitudes as the basic means of information processing. One set of objections arise from research on how pictorial imagery is processed in the brain (Cooper and Shepard 1973). Simply put, mental imagery does not follow the logic of sentences, and so a model committed to sentential attitudes does not cover this class of neuronal activity. This consideration feeds into hemispheric lateralization research which shows that the hemispheres process their 'data' differently: one in a verbal code, the other in an imagistic code (Sperry 1974). Now, for evolutionary psychologists and sociobiologists, the most intriguing objection lies in the phylogenetic continuity between language bearing species and non-language bearing species. To model all representations on sentential attitudes seems to call for a catastrophic break between hominid cognition and non-human primate cognition. Many evolutionarily committed thinkers such as Pinker (1994) are willing to side with the propositional-attitude camp here. (Pinker denies chimpanzees the *bona* fide 'language instinct.) But Churchland raises the objection that even if speech is a readily admitted addition to the hominid cognitive repertoire, many non-sentential forms of processing must still persevere. (She refers explicitly to phylogenetically 'old' brain structures which are involved in emotional behavior; these will be the focus of Chap. 2.) In my opinion, P.S. Churchland reasonably concludes that:

> Evolutionary and neurobiological considerations cry out against the likelihood that all representations in the brain are to be modeled on linguistic representations.... The implication... that vervet or rattlesnake brains manipulate sentences, noun phrases, verb phrases, and the rest is grimly farfetched though representations they surely do enjoy. (1983, 359)

There is a tendency amongst philosophers to slip back and forth between the idioms of rule-governed and mechanistic language. Sterelny (2003) provides the following examples that illustrate this point. Sterelny characterizes the parenting options of reed warblers, whose efforts are exploited by cuckoos who leave their own eggs in the warbler nest for the latter to hatch as if these were their own, as involving rules:

1. Treat all chicks in the nest as her own.
2. Treat all chicks as interlopers.
3. Try to discriminate between her own and the interloper and expel the latter.

Sterelny characterize these options as "learning rules". Sterelny explains that some species opt for the first rule rather than the third rule, reed warblers being a species that has a "learning rule" that sometimes permits exploitation (Sterelny 2003, pp. 12–13). The evolutionary choice of "learning rule" is discussed in terms of costs versus benefits that explain why one "learning rule" evolved for one species and a different rule evolved for a different species. But why pose the evolved identification and expulsion procedures or their lack specifically as a set of "learning rules"? Sterelny himself a few pages later characterizes cockroach escape mechanisms and firefly mate detection mechanisms without the aid of the rule metaphor (Sterelny 2003, pp. 14–15). Perhaps Sterelny could have availed himself of Dennett's 'as if' intentionality as a casual short-cut for the warbler-cuckoo example, but it should be clear that the appeal to 'rules' is superfluous when chemical and neural mechanisms account for the behavior. The rule model sneaks into the argument when characterizing the specification of the history of the evolutionary selection pressures that selected these mechanisms. What we need in this explanatory frame is a history of evolutionary events, not a system of rules and representations. In other words, the notion that rules and representations operate on top of, or in addition to, brain processes appears to be unjustified.

The question to pose at this point is whether the sentential model is an accurate representation of verbal behavior itself, even for language bearing species such as ourselves. Just because we produce linguistic behavior, is it necessarily the case that the brain processes that produce language are a mirror image of the behavior that is produced? After all, the brain is an electro-chemical system and there is nothing, unless you are a dualist, to justify the belief that the mind is a software program that runs on top of the brain's wetware in the form of sentential attitudes and their logical operations. There are several considerations that back skepticism on this point.

The first skeptical consideration is offered by John Searle. Searle makes no bones about it: the problem with Chomsky's rule governed linguistic universe lies in its lack of biological reality

> Precisely to the extent that we take Chomsky's analogy with biology seriously, to that very extent we do not need the hypothesis of rules and representations. If we are to think of the language organ on analogy with the heart or the liver, then we do not need the additional hypothesis of rule-governed behavior at all. The heart does not follow rules, nor does the liver. (Searle 1980a, 37)

Chomsky claims that the features of all existing natural languages are constrained by rules that permit only certain types of linguistic organization. Searle casts doubt on this thesis by way of making an analogy with visual perception. He argues that our inability to perceive infrared and ultraviolet light is not due to a "visual grammar system" that forbids it, but our inability is due to the structure and physiology of the brain. If the forms that language can take are similarly limited, this is also due to limitations of structure and physiology (Searle 1980a, p. 38).

Nonetheless, it is still possible that Chomsky's rules and representations model could be translated into the terms of a theoretical paradigm that has realistic neurological coordinates. Paul Churchland provides just such a possibility. Following the tensor network approach of Pellionisz and Llinas (1985) that has successfully modeled motor control, sensorimotor coordination, and sensory discrimination, Churchland suggests that their geometrical theory of phase-state transformation might also handle 'higher' cognitive activities. Without getting too detailed, Churchland speculates that something akin to the Chomskian model might be conceived along these lines:

> One might try to find, for example, a way of representing 'anglophone linguistic hyper-space' so that all grammatical sentences turn out to reside on a proprietary hypersurface within that hyperspace, with the logical relations between them reflected as spatial relations of some kind. (P.M Churchland 1989, 109)

Churchland makes it clear that this scenario is purely hypothetical at this point, and that it would require extensive research to back it up, but the philosophical point is that we need to pursue the development of a biologically realistic model instead of one based on the sentential attitudes of folk psychology (cf. Churchland 2007, pp.232–38). Should biologically based models succeed, it might be possible to claim that Chomsky's particular sentential model could be successfully reduced to tensor network theory. However, since Searle (1980a) points out that the rule-based paradigm is biologically unrealistic and incorrigibly so, it might be better to bet on theoretical replacement instead of reduction (Churchland 1989, p. 109, 1982, 1987).

There are, however, additional reasons to doubt that Chomsky's picture of language as rules and representations is satisfactory. Deacon (1997) has provided a picture of language acquisition that makes sense out of Searle's assertion that the range of possible linguistic organizations is a product of constraints present in the human mind but which are not based on rules. The other part of the language acquisition equation has to do with the fact language is a social phenomenon that exists outside in the practices of a verbal community. What exists outside the language learners' brain is public language. Deacon argues that the constraint on linguistic organization occurs due to the limitations of children's intelligence and the fact that they have to make sense of the practices of their verbal community in order to reproduce it successfully (Deacon 1997, pp. 103–110). That is, language must be adapted to the child's brain rather than the other way around. The constraints that emerge are produced by the simplifying assumptions that are necessary in practice to facilitate communication and reproduce the parent language. Deacon further argues that Bickerton's (1990) analysis of pidgins and creoles supports this view rather than the Chomskian view (Deacon 1997, pp 138–139; cf. Quine 2008, pp 186–187). Hence, the child does not come preprogrammed with a manual full of complex rules for translating the 'deep' mentalese structure of language in general into the specific language they learn, although the evolution of the hominid brain certainly does constrain the possible forms that all languages can take. Languages 'evolve' quickly in historical time as Deacon points out, and so the processes that enable successful acquisition and reproduction have to be on a very short lease of innate human cognitive constraints.

I think an important conclusion emerges from Deacon's analysis for purposes of this present essay. Cognitive significance is likely enjoyed without the vast apparatus of cognitive content hypothesized by the rules and representations model. Following Wittgenstein (1953), Wettstein (1988, 2004) argues that the significance of a piece of language lies not in its being an expression of innate meaning within a given human subject's mental representation, but is instead a function of its practical role within the linguistic practice of a community of speakers (Wettstein 2004, pp. 8–9). That is, the primary function of language is communication not representation.

Michael Tomasello (2008) also follows Wittgenstein's lead. He develops a probing account of the ontogeny and phylogeny of human communication that brings out the central role that background and contextual factors play in communication. The difference between great ape and human communication lies in the evolution of what Tomasello calls 'shared intentionality'. The latter refers to the ability to see that others can attend to that which we see and can respond to it collaboratively, and vice versa (see 2008, 320–327 for a summary of his argument). Great apes only evince individual intentionality. Their

cooperative efforts are based exclusively on their own self-interests, the cooperation that results being an artifact of the fact that they both want the same
thing and need to do the same thing in order to obtain it. Tomasello describes his
research that shows that humans are cognitively capable of and emotionally
motivated to help others in an altruistic manner. He argues that this is the result
of selection pressures on human cooperation in ancestral environments. The
more ancestral hominids came to depend on social cooperation, the greater their
social intentionality evolved. These abilities originated in an elaboration of
gestural forms of communicative competence that are derived from similar
abilities in our hominine forebears. Eventually, gestural forms of communication
were extended to verbal communication. The latter vastly enhanced the scope
and effectiveness of communication. The point that I would like to emphasize is
that Tomasello argues that much of what we think of as constituting the structure
of communication exists in the public sphere of the process of communication,
rather than in some sort of private internal sentential algorithm as hypothesized
by Chomsky and his associates. Although Hauser, Chomsky, and Fitch (2002)
attempt to parse the language faculty into two components, one being the
broader sort of communicative competencies described by Tomasello, they claim
that there is also a narrow, strictly linguistic, competency embodied in the
recursive 'rules' of universal grammar. Tomasello argues against this ur-gram-
mar model that the recursive features of linguistic communication reside in
shared intentionality. As a result, far from linguistic output being an expression
of internal sentential processes, it is more likely that abstract analytic models of
internal sentential processes are projections backwards of actual existing public
linguistic performances that take place in a given verbal community. Hence, as
Tomasello suggests, the rules and representations model places the internal cart
before the external horse (2008, p. 11).

Propositional-Attitudes and Mind/Body Dualism

Quine (1985) asserts with some confidence that the mind/body question has been
resolved in favor of the materialist position, but he warns that too loose a leash
on mentalistic talk will compromise this awareness. If we all agree that mental
events are just the behavior of the brain, what harm is there if we use the
mentalistic idiom for convenience if nothing else? Well, one danger is that some
of us might begin to think that mental events have nothing to do with the brain.
Jerry Fodor provides just such an example.[11] Others who hope to stay in good
stead with philosophical materialism, but who accept the existence of mental

---

[11] Fodor gave at talk at Rutgers University in February 2000 prior to the release of his *The Mind
Doesn't Work that Way* (Fodor 2000), where he asserted casually but with conviction that the
"brain has nothing to do with the mind". I was there; I heard it with my own ears.

processes, attempt to figure out how mental events are related to brain events. The problem is that the latter view assumes that dualism is true. If indeed there are mental events in a non-physical sense (Feyerabend 1963a, b) and the mind itself is non-physical, does it make sense to speak of the evolution of mental processes since non-physical mental events could not possibly be subject to physical laws? And, if that were the case, how could mental states or processes be said to be within the explanatory province of evolution? If the mind and its activities are non-physical, then we evolutionary psychologists are out of business and the dualists and Platonists will definitely be in business. Thus, the stakes behoove us to make sure that we construe mental life as a purely physical phenomenon. Churchland (1986) and Churchland (2007) admit philosophers such as Dennett (1991) into the materialist camp, but, even today, the majority of cognitive theorists and philosophers of mind (e.g. Clark 2008; Kim 2007; Shoemaker 2007; Block 2007; Pinker 2007; Minsky 2007) accept some version of an anti-reductionist program.

How is the "mind" related to the brain? Or is it? Are the mind and the brain two things or one thing? In *Neurophilosophy* (1986), Patricia Churchland provided an incisive critique of dualist perspectives in both folk psychology and scientific functionalist psychology (see also Churchland 1980a, b). Folk psychology refers to the commonsense use of terms involving beliefs, desires, perceptions, expectations, goals, etc. that are used to explain and predict behavior. One mental state is connected to one or more other mental states and these to a person's actions (Churchland 1986, p. 299). Substance dualists believe that the terms of folk psychology refer to non-physical mental states. On this view no reduction of the mental to the physical is possible. Nevertheless, following Descartes, Popper and Eccles (1977) held that the mental and the physical interacted in order to produce behavior. Physical sensations would present to higher-level mental perceptions and the thinker would use the physical sensory data in its reasoning and decision-making activities. One major embarrassment for this view was that neurological interventions via brain lesions or drugs affected the postulated higher-order mental events in ways that should have been impossible if the mind is genuinely independent from the brain (Churchland 1986, p. 319). An additional object that might appeal specifically to evolutionists concerns the evolution of a non-physical mind from lower species that only have physical components of mental life (see also Churchland 1980a).

A more promising view was advanced that proposed that the mind has properties that are 'emergent' with respect to the brain. This view is called property dualism. On this view, the mind is admitted to *be* the brain but subjective experiences are held to have different, and inherently non-physical, properties that the brain does not have. That is, at the brain level, we can measure the electrochemical activities, but these do not seem to capture the subjective experience of the reporting subject since these experiences seem to consist of *qualia* (i.e. properties such as redness, loudness, brightness, etc.). Both Smart (1959) and Feyerabend (1963a) have pointed out that just because qualia *appear* to have characteristics different from neuronal processes, it does not necessarily mean that they *are* different. Following in the footsteps of Smart (1959), the Churchlands compare cases in the history of science where one theory is reduced to a second theory such that the phenomena proposed by the first theory are shown to be examples of the phenomena proposed by the reducing theory. An example of this sort of inter-theoretic reduction is found in the reduction of 'light' to 'electro-magnetic radiation' (Churchland 1986, p. 324). It turns out that light *is* electro-magnetic radiation. Paul Churchland further illustrated the issue of illusory emergent properties in his essay, "The Betty Crocker Theory of Consciousness", by means of elucidating the relationship between 'heat' (i.e. the temperature of gases) and 'mean molecular kinetic energy'. In the Betty Crocker cookbook, heat is presented as being emergent with respect to the movement of the constituent molecules such that molecular movement causes something different i.e. heat to emerge, whereas, in fact, heat just *is* the mean kinetic energy of the interacting molecules (Churchland and Churchland 1998, pp. 113–122).

One begins to suspect that perhaps mental events might be nothing other than the behavior of the brain, and that the distinction between the two is based on a confusion of things and their activities. Perhaps there is a duplication of entities where in fact there is only one entity. The idea that the brain and the mind are the same thing is called monism. Quine has stated the case for monism concisely and with simplicity:

> Unless a case is to be made for disembodied spirits, we can argue that a dualism of mind and body is an idle redundancy. Corresponding to every mental state, however fleeting or however remotely intellectual, the dualist is bound to admit the existence of a bodily state that obtains when and only when the mental one obtains. The bodily state is trivially specifiable in the dualist's own terms, simply as the state accompanying a mind that is in that mental state. Instead of ascribing the one state to the mind, then, we may equivalently ascribe the other to the body. The mind goes by the board, and will not be missed. (1985, 5)

And yet we still have mentalist vocabulary and physical vocabulary and these two views appear to divide the world up differently and irreconcilably. It *appears* that the 'meanings' of the terms in the mentalist vocabulary can not be reduced to the physical referents in the materialist vocabulary. With respect to this issue, Quine refers us to Donald Davidson's notion of *anomalous monism* (Davidson 1963). Anomalous monism recognizes that every mental event is a physical event, but acknowledges that the vocabularies used to describe these events will diverge depending on the point of view.

There are a number of different paths that have been taken here. The Churchlands offer an alternative they call *eliminative materialism* where the old concepts of folk psychology can be replaced by new scientific terminology of our own invention (Churchland 1981, 2007; Churchland 1986). Stich (1996), on the other hand, allows that we can utilize the old terms of folk psychology but give them new referents in the new scientific psychology. However, many others (McLaughlin 1985; Chalmers 1996) choose to re-introduce some form of property dualism instead, thus transforming Davidson's anomalous monism into *anomalous dualism*. McLaughlin's 'defense' of anomalous monism achieves the following paradoxical conclusion: while "every mental event is a physical event"

(McLaughlin 1985, p. 333), it is also the case that the mental predicates do not have the same extension or reference as do the physical predicates. How can this be so? According to the functionalist, this is because mental states must be *functionally* defined over and above the physical states that bear or are said to instantiate them. John Searle (1984) illustrates the thesis as follows: we have many different concepts of ideas such as 'money' or 'marriage' in many different languages, and although each different term refers to the same concept, the underlying neurophysiological details are different in each case. Hence, there can be no hope of reducing or even correlating propositional meanings at the mental level of description to neurophysiological events.

Functionalists accept the argument that intentional states can be realized in a variety of physical codes, not only in the brain's wetware, but also in silicone chips, as well. Thus we see the intuitive appeal of the computer model for mental life: the mind is a software program that runs on top of the brain's neuronal hardware. This view has resulted in some strange formulations of mental life that attribute intentional states to a motley assortment of physical things. For example, Chalmers (1996) and McCarthy (1979) argue that thermostats have beliefs (i.e. it is too warm, it is too cold, or that it is just right). McCarthy claims that thermos containers are even more intelligent because they also know how to maintain the correct temperature for their contents (Searle 1984).

In general, however, functionalists claim that they are materialists in the sense that the mental event requires some form of material event in which to be embodied, but they also believe that the mental event is irreducible to the physical (Kim 2007; Block 2007; Shoemaker 2007; Clark 2008). Thus, mind–body dualism re-emerges in a new guise. Frankly, I think the problem lies within the language we use to discuss 'mental' events. Even the *Identity* hypothesis itself (Rorty 1965), which holds that all mental events are physical events, implies a fundamental dualism of mind versus brain. Feyerabend demonstrated this as follows: to claim simultaneously that x is a mental process of type A and also a physical process of type a, such that the two are identical, implies that mental events have physical features and that physical brain events have non-physical features (Feyerabend 1963b, p. 295). Feyerabend goes on to point out that if the identity hypothesis is correct then dualism is true, but also that if monism is true, then the identity hypothesis is false and that there are no mental processes in a non-physical sense. The problem appears to lie in the fact that when we say that two things are the *same*, there are still two things, not one. This, however, depends of what we mean by 'identity'. Place (1956) and Smart (1959), I think, successfully defend the identity thesis by restricting it to refer to identity in the 'strict' sense of the term, which means simply that lightning *is* an electrical discharge or that the *evening* star *is* the *morning* star—because both terms refer to the planet Venus. It is the same use of identity that we found in P.M. Churchland's identification of *heat* with *mean kinetic energy*. Hence, while our phenomenal experience, which we know by direct acquaintance, appears to be distinct from neuronal events, which will not be experienced phenomenally, it does not undermine the theory that that experience just *is* those neuronal events.[12] If we have the choice of two descriptions of a 'mental' event, one in terms of qualia and the other in terms of neural activity, which one should we prefer? After all, morning star and evening star are equally valid references to Venus. When it comes to matters of scientific explanation, Jack Smart thinks scientists should place their bets on the scientific, rather than the phenomenal (i.e. folk-psychological), alternative.

John Searle appears to profess acceptance of the physical stance in his concept of intrinsic intentionality. In chapter one of *Minds, Brains and Science*, Searle attempts to undermine mind–body dualism. He proposes that "mental phenomena, all mental phenomena, whether conscious or unconscious, visual or auditory, pains, tickles, itches, thoughts, indeed, all our mental life, are caused by processes going on in the brain" (Searle 1984, p. 18). Now, this passage all by itself without further qualification might be interpreted to suggest that mental phenomena are caused by physical phenomena but are themselves non-physical. Yet Searle adds a necessary proviso: "Pains and other mental phenomena just are features of the brain (and perhaps the rest of the central nervous system)" (Searle 1984, p. 19). Thus Searle appears to endorse the view advanced by Feyeraband (1963a, b) that there are no mental events in a non-physical sense. Searle's proposal and proviso thus appears to be in keeping with materialsim. Later in the book, however, Searle accepts the multiple instantiation thesis as described several paragraphs above (Searle 1984, p. 80). What could explain such a flagrant inconsistency?

I think the problem in the relationship between the mental and the material is at base terminological and not ontological. Our language for mental predicates does not match our language for brain events, but this does not mean that there are mental events in a non-physical sense (Feyerabend 1963a,b). As Quine (1985) points out, the lesson of anomalous dualism is this: mentalist language is used to categorize our experiences, and as a result, nominal groupings are formed that classify diverse things. These groupings do not necessarily cut nature at it joints. They do not form a natural kind. Thus our 'mental' representations consist of physical brain events, but these can be categorized in a variety of ways that do not accurately represent the physical details. As Quine surmised, a given concept is like a set of bushes, each of which has the same overall shape, but each of which varies in terms of its minute branching patterns (cf. Churchland 2007, 34). For example, there are 104 viruses that we label under the description 'cold'. The 104 viruses are real and distinct biological entities. We classify them under one nominal category: 'cold'. The functionalist, on the other hand, would claim that a

---

[12] Place referred to the thesis that phenomenal states are not identical to physiological states as the "Phenomenological Fallacy" (Place 1956, pp. 49–50). Hence, back in the 1950s, dualists raised the same objections to phyiscalism that non-reductive materialists (e.g. Kim 2007) raise today. Hence the rebuttals provided by Smart (1959) to those objections should still appeal to the physically inclined. I suggest that those readers who reject phyiscalist monism read Smart's ancient objections.

'cold' is an abstract essence and that this essence is implemented in 104 different material substrates. The functionalist would also claim that the 'cold' must be implemented in one of those 104 specific substrates, but they would deny that the cold is reducible to its physical substrates because it can be realized in so many different substrates. This is the same exact picture that functionalists give us of every mental predicate we could supply. The problem is that they have created an extra ontological level where only a nominal classification is in evidence. Thus, only an out-and-out Platonist or dualist can deny that mental events are, and must be, physical events in the brain. We can therefore accept a revised version of Towne's hypothesis:

> That strictly causal brain-processes may be found to underlie all rational thought processes, *including* the discovery that strictly causal brain-processes underlie all rational thought-processes. (Toulmin 1970, 25)

This thesis can be restated such that causal brain processes do not merely underlie rational thought processes but, instead, those rational thought processes *are* brain states. But, with this proviso in place, do rational thought processes play the central role their adherents claim they do in the causation of behavior? Davidson (1963, 1980) says, "yes", while Toulmin (1970) says, "no". Although Toulmin accepts Townes' hypothesis, he refers to reason explanation and causal explanation as ships passing blindly in the night. He also refers to the conflation of the two as a philosophical barbarism. Why? It might be because *reasons* invoke *intensions* (i.e. meanings) and thus fall outside the domain of scientific explanation (see section 'Folk Psychology: The Best or Worst of Intensions?'). Nevertheless, folk psychologists stake their claim on the causal status of intentions.

Intrinsic Intentionality and Folk Psychology

The reader may have noticed that I expressed no quarrel with Tomasello's use of 'intentionality'. That is because Tomasello uses the word in accordance with the commonsense concept of intentionality as a disposition to do something deliberately and the ability by another to see that one *intends* to do that thing deliberately. Because this form of intentionality involves the understanding of the intent by another, such that it is mutual, Tomasello prefers to call it 'social intentionality'. This mutual understanding, of course, can be recursive, and it can be so without the imputation that that understanding need necessarily be ensconced within a 'language-of-thought' or some other set of full blown beliefs (Sterelny 2003, p. 95). Because intentionality is conceived as mutual understanding that is rooted in non-verbal communication, we can go on to ask whether the achievement of *intrinsic* intentionality is sufficient to explain behavior since such intentions appear to require propositional-attitudes. Is an intention a self-sufficient cause of our behavior? Searle (1980a, b, 1984) seems to think so.

Despite Searle's failure to maintain the physical stance as revealed in our analysis of his commitment to the multiple realization hypothesis, he generally strives to be a materialist, and this comes across most clearly in his discussion of *intrinsic* intentionality. Searle (1980a) claims that the brain produces intentions just as the stomach digests pizza. In other words, intentions are simply physical activities of the brain. The question then becomes: how do intentions cause our behavior? According to Searle's notion of intrinsic intentionality, 'doing x' can not be explained apart from the agent's 'intention to do x'. The 'intention to do x' is constitutive of 'doing x'.

> The explanation of an action must have the same content as was in the person's head when he performed the action or when he reasoned toward his intention perform the action. If the explanation is really explanatory, the content that causes behavior by way intentional causation must be identical with content in the explanation of the behavior. (Searle 1984, 67)

Searle claims that a purely mechanical, physical, behavioral description of taking a walk, say, to Hyde Park is indistinguishable from walking towards Patagonia. What makes the difference is that the actor has intent to do one or the other and can tell us. The agent's telling us his intent to walk to either Hyde Park or Patagonia thus explains his walking to one or the other of those places and is the necessary and sufficient causal explanation of the action. Searle wants to argue that mental states that lack intentionality can not be credited with explaining behavior because they lack the type of agent generated preferred description intrinsic to the intentional state.

Does this brand of intrinsic intentionality provide a *bona fide* example of what evolutionary psychologists want for a proximate mechanism? Or, if his account is true, is it the case that evolutionary explanations are superfluous? The latter possibility may come as a surprise since many evolutionary psychologists and their philosophical advocates hold that beliefs and desires are *bona fide* proximate mechanisms. In fact, Searle makes it clear that he believes that all forms of sociobiology, behaviorism, and cognitive science are attempts to fill in the gap between the mind and the brain, and he believes by showing that the mind and the brain are the same thing that all these gap filling sciences will then disappear.

It seems odd, even at first glance that Searle should refer to behaviorism as a 'gap-filler'. In his paper, "Why I am not a Cognitive Psychologist", Skinner (1977) argued that our intentional states, when we have them, are explicitly recognizable as verbal behavior (i.e. speech) and that unspoken thought can be analyzed in terms of covert speech. Here, and also in "The Operational Analysis of Psychological Terms" (1948/1984), Skinner shows how privately received stimuli (i.e. physical sensations and perceptions) come to be linked to linguistic terms under the reinforcement contingencies supplied by a verbal community. Skinner's account is an historical analysis of how a speaker in a verbal community comes to attach language to his or her private experiences in such a way that members of that community can understand each other. Hence, Skinner does not provide a 'gap-filler'; instead he provides an analysis of how our private events or intentional states are conditioned by external features of the environment, in this case our social environment.

There have been other attempts to relate intentions to the social context in which they are expressed. The Action philosophy of Davidson (1963, 1980) appears to provide a case in point. However, the action philosopher is content to determine the criteria of 'appropriateness' for engaging in the action. Despite the attention to context of an action, Davidson's type of causal explanation explains people's acts strictly by reference to the contemporaneous and antecedent mental contents that accompany these acts (Davidson 1963, p. 685; Searle 1984, p. 87). The relevance of the context must explicitly be included in the mental content of the rationalizing reason. Hence we are still saddled with a view of explanation that focuses on the initiating role of reason. Prado (1981) has argued that the mere *initiation* of an action is an insufficient account of the cause of a given action. The lack of a causal analysis of the environment outside the skin commits the error of *psychologism*, in that it is imperative in giving a causal account of an organism's behavior to specify the various external contingencies of reinforcement that affect its behavior. What then of sociobiology or it is currently dominant allele, evolutionary psychology? After all, evolutionary analysis has as its object the delineation of the selective pressures in ancestral environments that led to the evolution of the motivational endowments that constitute the proximate mechanisms that are implicated in the causal explanation of behavior. Just as operant conditioning explains some aspects of ontogenetic behavior, evolutionary psychology explains crucial aspects of the phylogeny of our behavior. Each specifies the different modes of environmental selection that shape behavior either in the past or in the present, respectively. When we broaden our analysis to include both ontogenetic and phylogenetic environmental contexts, the psychologistic nature and inherent solipsism of folk psychological explanation becomes evident. Behaviorism and evolutionary psychology are not 'gap-fillers'; they provide theoretical perspectives that are necessary for the satisfactory explanation of behavior.

Let me provide an example of how an evolutionary psychology explanation and a folk psychology explanation might differ from one another. The folk psychological explanation will focus on the immediate intentions of an actor, while the evolutionary psychological explanation will point at the evolutionary history that explains why those particular dispositions, rather than some other set of dispositions, are found to operate in some given set of circumstances. Let us say, for example, that Donald desires to have a homosexual encounter. According to the folk psychologist, who is committed to intrinsic intentionality, it is sufficient that we know that this is what Donald desires. However, to get the whole story, we also need to know, that having that *desire* and *believing* that the basement restroom at the *Roxy* on 11th and Stark might afford such an opportunity, that Donald takes the bus downtown, enters the *Roxy* and heads for the basement restroom with the intent of finding a partner, and that he can tell us that if we ask. Belief + desire = intention. What more could you want? Well, if you are an evolutionary psychologist you might want to know why Donald wants to do such a thing in the first place. You might wish to enhance the folk psychological explanations with theories of motivation that point to the operation of evolutionary factors. You might, for example, hypothesize that Donald prefers to seek a homosexual encounter rather than a heterosexual encounter in terms derived from sexual strategies theory

(Symons 1979). The optimal male strategy is to seek low cost promiscuous sex while the optimal female strategy is to demand signs of commitment prior to sex and thus to play hard to get. Current stimulus conditions can be added to show how such a strategy might emerge in a given set of circumstance. Say, based on his prior experience with trying to attract women partners, Donald has come to believe that the costs outnumber the rewards of pursuing female partners. Perhaps, also on the basis of experience, or perhaps he has heard of it in connection with stories he has heard from others, Donald also believes that homosexual encounters are more casual, thus demanding less effort and fewer demonstrations of commitment or resource holding potential. I would agree that these considerations make for a better explanation than a mere intrinsic intentionality explanation.

The point here is that intrinsic intentionality does not get us very far in the explanation of Donald's behavior. It is certainly inadequate to give us a picture of the ultimate or proximate causation of his behavior. We also have to question the role that beliefs and desires play in the explanation. In an evolutionary explanation that specifies the evolutionary factors and proximate mechanisms that underlie the behavior, the beliefs and desires themselves become behavior that requires explanation. Even though I employed the term 'strategy' above in my explanation of Donald's behavior, I do not intend this term to be understood as either a conscious 'strategy' or an unconscious 'strategy' or a program that is running in Donald's brain. Here 'strategy' refers to the evolutionary history of the events that led to the fixation of the disposition to seek low-cost sexual encounters, which, according to Symons, is an adaptation of heterosexual male psychology that has been hijacked because the homosexual encounter offers the exercise of the pure male 'strategy', while heterosexual encounters require the male to compromise with the female 'strategy', which favors monogamy and demands proof of fidelity *viz a viz* mating effort. It is difficult not to use the belief-desire terminology of folk psychology, even in evolutionary explanations. The necessary step is to conceive beliefs and desires as verbal behavior that needs to be explained, rather than as causal mechanisms that explain that behavior. In addition to specifying the evolutionary history of a given 'strategy, we also should strive to get at the proximate mechanisms that produce such behavioral dispositions. Hence we need to identify the brain mechanisms in the hypothalamus and in other potentially relevant brain structures. The evolutionary 'strategy' provides the history of the adaptation. We need to identify the adaptation in order to link it with the proximate mechanism, that is to say, the actual physically existing brain mechanism that produces the behavior or the disposition to engage in the behavior.

Folk Psychology: The Best or Worst of Intensions?

I asserted above that *beliefs* and *desires* are *behavior* rather than proximate mechanisms. That move is designed to undercut their status as explanatory forces. On the other hand, Dan Dennett appeals to beliefs and desires in order to defeat behaviorism. In *Kinds of* Minds, he relates the following quip about behaviorists: "they do not believe in beliefs, they think that nothing can think, and in their opinion nobody has opinions" (1996, p. 120). Now, I believe that Skinner would think that was pretty funny. And that is not just my opinion because, in his autobiography, Skinner (1979, p. 80) grants that kind of talk the status of being a harmless way of speaking in our everyday language.[13] Nevertheless, Skinner thought that a science of behavior has to do away with mentalistic language. Both Skinner and Quine think that behaviorism is the cure for mentalistic explanation. Alternatively, coming from cognitive neurobiology, Paul and Patricia Churchland think that the concepts of folk psychology can be eliminated in favor of a computational theory of representation.

I think the first issue that needs to be addressed is what is it that we are referring to when we utilize mentalistic concepts such as 'mind' or 'belief'. Although, 'beliefs' are widely, if not ubiquitously, held to be a fundamental building block of folk psychology (Fodor 1975, 2000), there is reason to doubt that 'beliefs' refer to actually existing mental events. Quine and Ullian (1978, pp. 10–11; Quine 1987, pp. 18–21) regard beliefs to consist of dispositions but not to have the same ontological status as thoughts or representations. This is because beliefs form a heterogeneous grab-bag of entities. This bag includes many thoughts that have never occurred and never will to the person who is said to harbor them. For example, if you believe that your lamp is standing 3 feet away, do you also have the belief that it is not 3.1 feet away? Although you could assent to an infinite number of queries, it is absurd to think that all your potential responses are actual mental states that exist in your mind-brain. Hence, Quine and Ullian suggest we should not speak of 'beliefs' as mental states but as dispositions that can be made manifest under certain conditions.

When we do have actually occurring private events another problem presents itself. That is because in every case the exact neurological details vary from individual to individual as Quine (1985) indicates. The unifying factor is that the behavior that results is defined in public language that gives it a uniform shape. Recall Quine's metaphor: although the individual branching of the twigs on two separate bushes may vary, the overall shape of the two bushes is trimmed to the same shape. This is why Quine opts for behaviorism over neurological reduction in the explanation of behavior. Behaviorism provides an extensional account of behavior (Quine 2008). By focusing on the public extensional realm the problem posed by the internal heterogeneity of representations is avoided. The problem here is that of *intensionality* (as distinct from *intentionality*). Intensionality with an s refers to *meaning*. Although the evening star and the morning star both refer to the planet Venus, the different intensional terms individuate different senses by which the referred to object is identified (Dennett 1996, pp. 38–40). For example, my intensional concept of Vienna is idiosyncratic to my experiences of Vienna,

---

[13] Research on 'Theory of Mind' by Gelman (e.g. Gelman et al. 1994; Gelman 2003) suggests that we are naturally inclined to think of ourselves and others in terms of psychological essentialism; however, this issue must be distinguished from the issue of metaphysical essentialism.

which will necessarily be different from some other individual's, since our experiences differ. Each of our intensional bushes is highly individual and hence, according to Quine, not suitable for reductive explanation. The psychobiologist, Uttal (2005), agrees and concludes, based on the highly intricate neurological bushes that necessarily obtain in individual cases, that a reductive cognitive neurobiology of mental representation is impossible. He calls this the *intractability* problem. Thus we can see that the appeal of philosophical intentionality, the language of thought paradigm, and the propositional-attitude psychology that supports it is inextricably committed to Platonic idealism. The only way out of the internal quagmire of messy *intensional* states for the mentalist is to define them as noise in the intentional system. Materialists, such as Uttal, therefore opt for behaviorism as providing the best bet for scientific psychology. Quine, on the other hand, is cautiously optimistic about the potential that reductive cognitive neurobiology shows for reclaiming at least some of the territory that is currently the property of the *intensional* idiom of folk psychology (Quine 1990, p. 72).[14]

Conclusion: The Mechanist Stance

What is the lesson of the preceding analysis? In what capacity can evolutionary psychology and the neurosciences hope to contribute to a science of human behavior? The answer is by focusing on adaptations and proximate mechanisms. I propose to call this perspective the *mechanist stance*. Although Dennett presents the physical and design stances as alternatives, it is clear in evolutionary explanation that these two stances are combined via ultimate and proximate levels of explanation. The Mechanist stance utilizes that complementary articulation. We can take the mechanist stance instead of the intentional stance, which introduces a Platonic bogeyman. This ghostly intermediary can safely be eliminated.

The goal of this excursion into propositional-attitude psychology is to put into question the adequacy of intentionality based models to serve evolutionary psychology with an account of adaptations or proximate mechanisms. To be sure, the proximate mechanism *is* a matter of the wetware of the brain, but adaptations are not computer software programs that run on the proximate wetware of the brain. Rather, if the evolutionary history of a proximate mechanism can be shown to involve adaptation, the proximate brain mechanism is an adaptation. Determining whether the brain mechanism is adaptive or not requires the specification of the selection pressures that produced it, but this history is also not a computer program. Thus, formulating our theories on secure mechanistic ground will ensure that we avoid the pitfalls and perils of the intentional program and the *mis-*

---

[14] If you prefer messy intensionality to science, then I recommend that you follow Derrida's lead into literary criticism. Alternatively, you could follow Wittgenstein into ordinary language philosophy.

representations provided by information processing models and 'as if' intentionality. Certainly, proximate mechanisms will feature both indexical systems that model the stimulus and imperative systems that motivate behavior. P.M. Churchland provides a version of the physical stance that accounts for cognition in terms of biology.

> More generally, the perspective on cognition that emerges from neuroanatomy and neurophysiology holds out an entirely novel conception of the brain's fundamental mode of *representation*. The proposed new unit of representation is the *pattern of activation-levels* across a large population of neurons (*not* the internal sentence in some 'language of thought'). And the new perspective holds out a correlatively novel conception of the brain's fundamental mode of *computation* as well. Specifically, the new unit of computation is the *transformation of one activation-pattern into a second activation-pattern* by forcing it through the vast matrix of synaptic connections that one neuronal population projects to another population (*not* the manipulation of sentences according to 'syntactic rules'. (2007, p. 23)

Churchland goes on to suggest that the anti-reductionist view espoused by cognitive functionalists has it backwards. If it appears that a given cognitive function can not be reduced to the brain's physical activities, this is because that purported function forms a false isolate—that is, several different things grouped as instantiations of one thing. I think this is why, in a review of connectionist theories of representation, Globus (1992) concluded that dualism was ubiquitous and unavoidable, and why, he rejects any and all forms of connectionism. Although I believe the extreme pessimism of critics such Uttal and Globus is unwarranted, I agree that 'representational' language tends consistently to invite the specter of mind/body dualism. Thus I believe that the most promising avenue in the study of proximate mechanisms lies in the area of motivational endowments because these are explicitly non-cognitive and because they play a fundamental role in learning and motivation. Once we come to appreciate the central role that motivation plays in the explanation of behavior we will also be in a better position to see where and how a cognitive theory of representation (or a theory of indicatives) fits into the explanatory picture. Hence, the issue of motivational endowments will be the focus of Chap. 2.

Chapter 2
Why I am not an Evolutionary Psychologist: On the Imperative Nature of Motivational Endowments

In Chap. 1, I attempted to show why propositional-attitudes take evolutionary psychology down wrong path. In this chapter, I intend to address the role that motivational endowments, or what Millikan (1993) refers to as *imperatives*, play in the explanation of behavior. Some philosophical psychologists have attempted to conceptualize motivational endowments in terms of intentionality. Sterelny (2003), for example, refers to motivational endowments as 'preference-like' states in a way that corresponds to his formulation of cognitive states as 'belief-like' states. The problem, of course, lies in the 'as if' nature of the formulation in both cases. Sterelny wishes to avoid both the Scylla of Fodor style language of thought psycho-semantics[15] and the Charybdis of Churchlandian neuro-computation, which eliminates language of thought psycho-semantics, but in doing so he has left us with 'as if' intentionality as the only alternative (p. 96). Although Millikan conceives her paradigm of cognitive *indicatives* and motivational *imperatives* in terms of *biosemantics*, that is to say adaptations (or what she calls 'proper functions'), she ends up deploying these in the idiom of 'as if' intentionality. I suggest that we let Jerry Fodor open the escape hatch for us. If we wish to forestall a commitment to propositional-attitude psychology, we potentially start out on the right foot because Jerry Fodor excludes emotions and other aspects of motivational endowments from the realm of propositional-attitudes.

Despite Fodor's ejection of motivational endowments from cognitive science, motivational endowments clearly perform a crucial role in the scientific study of behavior. If not a proposition based model, then what type of model can best help us understand the role of motivation in the explanation of behavior? Contrary to Fodor's a priori elimination of emotion from cognitive science, it may turn out to be the case that a neuro-computational model can account for aspects of emotional behavior as well as cognition. Cottrell and Metcalfe (1991) have developed a connectionist model that is capable of distinguishing the eight basic emotions

---

[15] Sterelny refers to this view as the 'Simple Coordination Thesis' (2003, p. 6).

A. Walter, *Evolutionary Psychology and the Propositional-attitudes*,
SpringerBriefs in Philosophy, DOI: 10.1007/978-94-007-2969-8_2,
© The Author(s) 2012

(see also Churchland 2007, pp. 95–96). Along these lines, Rolls (1999) believes that emotion may eventually be modeled in terms of tensor network-principles (319). Whether or not tensor-network analysis can be successfully applied to the study of emotion and motivation, there is strong reason to believe that the emphasis on representation over motivation has led to a serious neglect of the role motivational endowments in the explanations of behavior that are offered by evolutionary psychologists. Certainly, the omission of any treatment of the topic of emotion in David Buss's *Handbook of Evolutionary* Psychology (2005) provides evidence of such a lacuna. Panksepp and Panksepp (2000) make the issue of motivation central to their critique of evolutionary psychology. It is thus to the issue of motivational endowments that we now turn.

Motivational Endowments

Paul Churchland points out that cognitive activities, even across species, will share a similarity space because individuals, including individuals of different species, share, to varying extents, an evolutionary history that conserves neuronal structures and functions. Although Churchland makes this point specifically in regard to representational systems, Panksepp and Panksepp (2000) argue that evolutionary conservatism is especially salient in the case of the reward circuitry and motivational systems of the brain. Pfaff (1999) also argues this point:

> The inference here is that mechanisms of neural integration and hormonal control evident in reductionistic studies of lower animal sexual behavior must apply to the explanations of the biological side of human libido, the primitive physiological component of human sexual motivation. Because of the marked parallelisms in some features of primitive reproductive behaviors and striking conservation of the neural and endocrine mechanisms controlling them, we propose an extension of the tremendous body of work on animal neuroendocrine and sex behavior mechanisms to the biological understanding of human libido. (199)

To be sure, species differences do exist. Pfaff's research on sex drive is largely devoted to mapping the lordosis response in the female rat, and human females do not exhibit lordosis, partly because of species based differences in the connections between the motivational structures of the limbic system and the muscular-skeletal system. Hence, species differences in structure and function must be accounted for as well. Even among higher primates, both conservatism and derivation are evident. For example, with respect to primates, allometric scaling of limbic structures shows that the size of the limbic cortex in humans is similar to that of other primates. This leads Armstrong to conclude that the limbic cortex continues to perform the same motivational functions in humans that it does in monkeys and apes (Armstrong 1992, p. 123). Armstrong, however, notes several specific structures where humans deviate from apes. First, the olfactory structures in the limbic cortex are relatively smaller in humans compared to apes and old world monkeys, suggesting a decreased role for olfactory stimuli in human motivation. Second, the anterior thalamus is larger in humans compared to apes and old world monkeys. This suggests to Armstrong, given the connections between this structure and cortical association areas, that humans are better equipped than other primates to remember and associate the emotional significance of stimuli, including symbolic stimuli (Armstrong 1992, pp. 124–131).

Although structure size is certainly important in gauging evolutionary change in function, we must also ask about changes in organization. Differences in behavior between species must point to differences in their motivational endowments. Hence, although of fundamental importance, the study of gross anatomical differences between species cannot provide answers to questions of species-specific motivational dispositions. Research specifically devoted to the neural mechanisms of emotion and motivation in *Homo sapiens* must be addressed in order to make evolutionary psychology a progressive research program. Aldridge (2010) reviews evidence that demonstrates that the human prefrontal cortex is not only larger in humans than it is in apes but that it also is morphologically different, which suggests that significant reorganization of cortical and subcortical components occurred during the course of human evolution. Heatherton (2011) and Steklis and Lane (2012) further discuss the implications this has for social cognition in humans. Although functional studies are admittedly few, the changes in morphology of the prefrontal region in humans compared to great apes fits in with the received view that humans regulate their emotional responses by means of cognitive systems. Monitoring the emotional reactions of others as well as mastering the control of one's own responses and actions appears to depend heavily on the activation of the frontal pole (Brodmann's area 10) as revealed in neurophysiological studies conducted by Buckner et al. (2008) and Rilling et al. (2007), which demonstrate a shift toward frontal pole activity in humans compared to chimpanzees.[16] This has been interpreted by Steklis and Lane (2012) to support the claim that emotional regulation is guided by cognitive monitoring of self and others. Steklis and Lane argue that this shift in brain organization underlies the uniquely human capacity for what evolutionary psychologists have come call the 'Theory of Mind' (Leslie 1994).

Proximate Mechanisms: The Central Role of Emotion, Drive and Reward

We have seen that, under the stewardship of evolutionary psychology, proximate mechanisms have come to include a motley assortment of entities such as intentions, beliefs, desires, and strategies which are hypothesized to produce adaptively relevant behavior. The intentional idiom is fundamental to the goals of 'Theory of Mind' research since the latter attempts to explain the evolution of 'mind reading'

---

[16] Hence it turns out to be the frontal pole, not Dennett's "East Pole" that turns out to be the cognitive center (Dennett 1998).

(i.e. the ability to ascertain the motives and beliefs of others, and the ability to manage this task in recursive steps).[17] Despite the central importance that such states are accorded in the 'Theory of Mind' theory (ToMM), it is not clear, however, that even if such entities were to be conceived as physical brain events, that they should be classified as a theory of cognitive representation. In other words, the task of ToMM research will still be to explain the evolution of the ability to ascribe 'intentions' and 'beliefs' to actors but to do so without the further implication that such an account is also a theory of the structure and dynamics of representation or motivation as claimed by evolutionary psychologists such as Leslie (2011) who accept the tenets of folk psychology at face value. Theories of representation remain the province of cognitive neurobiology (e.g. Churchland and Chrchland 1998) and theories of motivation remain the province of affective neuroscience (e.g. Panksepp 1998). Evolutionary psychologists need to position their research in relation to these two latter disciplines instead of buying into the LOT model. The inclusion of intentional phenomena as members of the class of representational and motivational structure and dynamics is misleading and confuses the relationships that hold between brain, behavior and environment, as I intend to show below.

When we try to explain behavior, emotions, drives, and the reward circuits of the brain should obviously play a central role. Emotions may seem to have a foot in the physical door but also in the mentalist door since they also qualify as *qualia*. That is we have subjective experience of emotions as they occur. Recall, however, that according to Searle's *proviso* all properties of physical brain events are also physical events in the brain no matter how non-physical they appear to be. How then should we characterize emotions? According to one of the more noted researchers in the field, Joseph LeDoux, emotions did not evolve as conscious feelings. "They evolved as behavioral and physiological specializations, bodily responses controlled by the brain that allowed ancestral organisms to survive in hostile environments and procreate" (LeDoux 1996, p. 40). LeDoux, who sometimes appears to buy into the non-physical nature of cognition, goes on to assert that emotions, being of the body, are physical in nature. Thus, there is a tendency by researchers in this field to divide the mind-brain into the mental-cognitive part and the physical-emotional part. For example, Zajonc describes his two system theory as such:

> The limbic system that controls emotional reactions was there before we evolved language and our present form of thinking…. When nature has a direct and autonomous mechanism that functions efficiently… it does not make it indirect and dependent on a newly evolved function [cognition]. (1980, 169)

Since emotional systems evolved prior to language systems, it is clear that their method of processing must be in a form other than the language of thought

---

[17] That is to be able to understand not only that "I know what you are thinking, but rather that I know that you know what I'm thinking, or even that she knows that you know what I'm thinking", and so on.

(Panksepp and Panksepp 2000). Fodor argues that this why the study of emotions is not a suitable candidate for cognitive science (1975; see also LeDoux (1996, pp. 34–35). When discussing emotion, we must talk in terms of electro-chemical activities in neurons. Because emotional processes are non-verbal, they must be at least partly encapsulated from conscious volition. Research that contributes to this aspect of emotions consists in the famous studies of universal facial expressions described by Ekman and Friesen (1975) and Izard (1971). Although ways of classifying the basic expressions of emotion differ a bit, the polythetic cluster usually includes fear, anger, joy, sadness, disgust, and surprise. Ekman refers to these cognitive systems as 'automatic appraisal systems'. Recognition of these emotional expressions is automatic and obligate and largely impervious to cultural penetration. In other words, the cognitive processing of emotions appears to be *modular*. According to Fodor, however even if the recognition of emotions may be encapsulated, they are not modular in the same sense to which he ascribes encapsulation to perceptual processes because only the latter are involved in the fixing of beliefs (Fodor 1983).

Nevertheless, Zajonc (1980) argues further that the *production* of emotional expressions is also modular, as does Griffiths (1990; 1997, pp. 91–97).[18] However, cognitive recognition and expression systems may achieve *modularity* in fundamentally different ways, and a potentially important distinction needs to be drawn between them. Following Elman et al. (1996), Buller (2005) argues that the modularity that occurs in recognition or appraisal systems (what he calls "higher" cognition) is achieved ontogenetically by learning (pp. 128–143, 195–200). This is because specific neural systems in a maturing brain are established by the reduction or pruning of initially prolific connections. Buller calls this the principle of 'proliferate and prune' (Buller 2005, 134, 195–200). That is, representational modules are the *product* of experience rather than a pre-determined hard-wired set of structures as suggested by Tooby and Cosmides (1989, 2005). Buller argues that neural circuits are not genetically pre-determined (cf. Elman et al. 1996). In apparent agreement, Paul Churchland (2007, p. 43) describes how a partitioning of emotion categories might come about within a dynamic tensor network model that starts out from an initial blank slate position. This view places Buller and Churchland in what might be called the 'blank slate' camp, since experience in ontogeny provides the structure for what starts out as an equipotential system.

Opponents of the blank slate view such as Marcus (2004) argue that the development of the brain, or any other biological structure, is not genetically underdetermined and that it is wrong to think that the basic layout of the brain is primarily a function of experience. Marcus claims that the brain is not hard-wired but *prewired*. Developmental plasticity plays a fundamental role, but it is a guarantor of the innate structure of the brain—not its antithesis. As Marcus pithily

---

[18] One could therefore argue that the modularity of emotional expression is fundamental to the fixation of desire, that is, if one were inclined to defend a propositional-attitude approach to emotions.

puts it, "[n]ature provides a first draft, which experience then revises" (2004, p. 34). Marcus refers to this as the paradox of development. Flexibility and innateness both work together to produce brain organization. Flexibility provides a corrective mechanism that irons out errors in the developmental sequence. Marcus argues that anti-modularists such as Karmiloff-Smith (1992) should not interpret the lack of specific genetic markers that correspond to specific brain regions as evidence against genetic specification of brain systems. Marcus suggests instead that a combination of overlapping molecular markers specify the construction of specific neural systems (Marcus 2004, pp. 84–85). In other words, the brain may indeed be some kind of 'Swiss Army Knife' as proposed by Cosmides and Tooby 1994—but Marcus (2008) believes that it is likely to be more of a gerrymandered 'kluge' than a perfectly adapted system of modules as claimed by Cosmides and Tooby.

The division of labor between the cognitive and emotional systems is instructive. Representational systems have to be plastic in order to absorb and configure data about the external world. What occurs in the training of an emotion recognition phase-space network is the fine tuning of discriminations, including those that involve the linguistic categorization of emotions (Churchland 2007, Chap. 3). In contrast, emotional response systems, which reside in the phylogenetically older limbic cortex, are produced under more highly canalized control of genetically specified mechanisms, as openly admitted by critics of modularity such as Buller (2005, p. 143). These motivational systems are innately specified and are unlearned. The important point is that the emotional response circuits in the limbic cortex are less plastic than those for "higher" cognition in the cerebral cortex. Hence, elementary motivational endowments come pre-partitioned into modules by evolution, whereas representational systems are designed by evolution to be modified by experience. Perhaps more importantly, motivation and reward systems provide the basis for reinforcement upon which learning depends. If there is a genetic leash that tethers cognition and culture, it manifests itself in the reward, drive, and emotional mechanisms of the brain.

Are Emotions Proximate Mechanisms?

As discussed above, emotional responses are produced by brain mechanisms in what is frequently referred to as the limbic cortex. Fig. 2.1

What is expressed outside in behavior is also experienced inside. Hence emotions qualify as *qualia*. They are subjectively experienced. Magda Arnold relates familiar emotional *qualia* to the events of the drive and reward systems:

> Relaxation, well-being, pleasure, and positive emotions seem to be mediated via the serotonin system, modulated by enkephalins and endorphins; and pain, malaise, and negative emotions appear to be mediated by the nor-adrenaline system, modulated by substance P. (1984, 388)

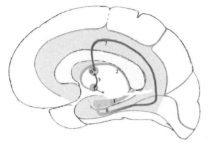

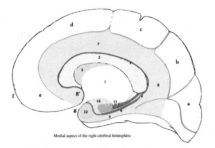

1 : Hippocampus / **2** : Fornix / **3** : Corpus mammilare /
4 : Tractus to Hypothalamus & Troncus Cerebri /
5 : Tractus mamillothalamicus / **6** : Nucleus Anterior Thalami /
7 : Cingulum / **8** : Tractus Cinguli

1 : Area of th thalamus / **2** : Septum lucidum / **3** : Corpus callosum /
4 : indusium griseum / **5** : Fasciola cinerea / **6** : Gyrus dentatus /
7 : Gyrus Cinguli / **8** : isthmus / **8'** : area sus callosa /
9 : Gyrus parahippocampal /**10** : uncus / **11** : Hippocampus / **12** Amygdala /

a : Cuneus / **b** : Pre-cuneus / **c** : Paracentral lobule / **d** : Gyrus Superior frontal /
e : Gyrus médial orbito-frontal / **f** : Frontal pole / **g** : Temporal pole

(a) 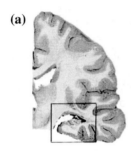 **(b)**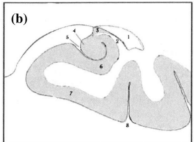

A: Coronal section of the right Cerebral Hemisphere / Limbic area
B: The area outlined in A is enlarged in B

1: Ventriculis lateralis / **2**: Hippocampus / **3**: Fimbria / **4**: Dental gyrus / **5**: Hippocampal sulcus
6: Subiculum / **7**: Parahippocampal gyrus / **8**: Colateral sulcus

Fig. 2.1 James Papez suggested the involvement of the limbic system in emotional expression in 1937. His hypothesis is based on the relationship that exists between the hypothalamus, the mammillary body, anterior thalamic nucleus, cingulate gyrus and hippocampus. Such a relationship normally constitutes a harmonious whole that is the center which elaborates the emotional function. In other words, it suggests the existence of an anatomical substratum of the emotional experience as thought and lived. The fornix is the main anatomical structure making the connection between the components of the circuit (Drawings and captions by Taya Alami, M.D.)

Arnold can make these assertions without violation of Searle's *proviso*.[19] The subjective feelings can be construed non-dualistically as physical properties of a physical system.

The other crucial feature specified in LeDoux's quote above is that emotions are adaptive. They exist as they do because they enabled survival and reproduction. Therefore, how can we best characterize the relationship between emotional

---

[19] See p. 21 in Chap. 1.

mechanisms and adaptation? One alternative is construe them in the manner offered by Griffiths, who prefers to conceptualize emotions as an *affect-program* system (Griffiths 1997). The affect-program initiates brief, highly stereotyped behavioral tropisms from hard-wired brain mechanism in the limbic cortex (1997, p. 100). Although the analytic tools employed by Griffiths are genetic, developmental, and adaptive in character, the choice of the term 'program' seems to imply agreement with the software analogy. In fact, Griffiths characterizes the 'poison-avoidance mechanism' in the rat that triggers food item rejection after illness as a set of "idiosyncratic rules" (Griffiths 1997, p. 108). Are we really expected to believe that the rat's mind-brain is a sentential engine that operates in terms of propositional-attitudes? Griffiths himself insists that affect-programs are reflex-like unconditioned responses (Griffiths 1997, p. 117), so it is difficult to accept the rationale behind the gratuitous ascription of the program metaphor to the explanation.

Leading proponents of the evolutionary psychology program, such as Tooby and Cosmides (1990), also exhibit a commitment to 'as if' intentionality when it comes to the explanation of emotions. Although Tooby and Cosmides regard emotions to be evolved proximate mechanisms, and their concern is to identify the pressures in ancestral environments that selected them, what is gained by characterizing cues that trigger emotional responses in terms of propositional-attitudes that characterize them in terms such as "looming approach of large, fanged animal" (Tooby and Cosmides 1990 quoted in Griffiths 1997, p. 116)? Griffiths himself points out that the "appraisal mechanisms of affect program states simply do not embody knowledge of the evolutionary past in this form" (1997, p. 116). Griffiths notes that "insofar as the mechanism reflects details of the evolutionary past, it does so in the form of learning preparedness" (Griffiths 1997, p. 116). That is, an emotion is *not* a propositional-attitude waiting to happen. Primary emotions are unconditioned responses to unconditioned stimuli and secondary emotions are conditioned responses to conditioned stimuli. All emotion systems are therefore systematically tied to the reward circuitry of the brain.

Frank (1988) makes an interesting point in connection with the social functions of emotions that ties them to the reward system. His insight is that emotional expressions function to communicate the state of their bearer to other social actors, and that this is useful both for the emotion bearer and the emotion perceiver in that it exerts control and counter-control over the outcomes of social interactions that may benefit both parties at both ends of the social exchange. That is, the emotional expression may encourage or discourage the actions of others in ways that may benefit social cooperation. Moreover, Frank argues that the irruptive nature of emotions serves to act as a circuit breaker that interferes with calculated self-interested behavior.

The latter point is important in that it sheds potential light on the role that cognition is argued to play in the regulation of emotion. Sterelny (2003) has argued that over the course of human evolution multifaceted and hostile environments selected for a decoupling of cognition from emotion. He argues that behavioral responses are not obligate as they are in reflex based systems—

Dennett's Pavlovian critters. Sterelny argues that cognition has turned the tables on motivation in the hominid line because reproductive success came to depend increasingly on a multivariate evaluation of relevant factors and potential responses (Sterelny 2003. pp. 92–96). The actor has to think about a number of possible response-consequence contingencies and choose the best. This way of thinking places the rational agent firmly in charge of his or her motivational endowments. But this how emotion-cognition circuits actually work? Damasio (1994) summarizes a large body of research on emotion that shows that damage to the brain's emotional circuits impairs the ability to make even simple decisions. Following Damasio, Churchland and Churchland (1998) made a convincing argument that behavioral decisions require an intact emotional system otherwise the reward value of the stimulus cannot be recognized as such and cannot guide behavior towards a goal. The power of social reinforcement as a reward or punishment certainly plays an important role in explaining its status as an important class of adaptations. Hence, emotions and the brain reward mechanisms that they depend upon appear to be driving the buggy, not cognition per se. These reward mechanisms themselves therefore need to be accounted for in connection with the adaptive nature of emotions. Certainly, one would think that a system as *prima facie* electro-chemical physical as the neural machinery involved in reward circuitry would be impervious to usurpation by propositional-attitudes. Think again.

The Ghost in the Reward Mechanism

During the long tenure of behaviorism it was difficult if not impossible to disentangle brain reward from brain drives within reinforcement theory. Both were linked to increases or decreases in respondent behavior, and interruptions in the functioning of any of the various neurophysiological structures that were implicated in the process interrupted the behavior (Olds 1977; Routtenberg 1978; Wise 1978). Morillo (1990, 1995) refers to models that focus on reinforced behavior as the "behavioristic learning theory of motivation (BLTM)" (Morillo 1990, p. 172). Although Morillo finds it emblematic (and problematic) that behavior was directed at obtaining external objects of satisfaction such as food, water, or sexual relations as pursued on the behaviorist program, these unconditioned stimuli and unconditioned responses remain at the center of the neurobiology of emotion (LeDoux 1996). Morillo has correctly identified the reward event as the fundamental process that anchors learning with respect to positive and negative reinforcement (1990, p. 176; 1995, Chap. 2). She calls this 'reward event theory (RET)'.

The key was to distinguish neural systems and their activities that drove behavior towards or away from a given stimulus from the reward event in the brain that occurred upon consuming or engaging the stimulus. Wise and Bozarth (1984) made some progress into the ability to draw this distinction by identifying the sequence of structures and events in the nervous system that culminated in the reward event. They identified a series of four elements that led to the reward event; the first elements of the series were the drive elements. The basic motivation expressway runs from the ventral tegmentum in the hypothalamus to the caudate nucleus which connects the hypothalamic structures to the cerebral cortex via the medial forebrain bundle and a host of supporting structures. The ascending and descending fibers of the medial forebrain bundle run from the ventral tegmentum to the caudate nucleus and house the dopamine and nor-epinephrine cells that provide the positive and negative push and pull drive processes towards or away from the stimulus.

Although initially there was confusion as to whether the reward event was serotonin based or endorphin based, it emerged that the triggering of endorphins was the reward event and that serotonin provided the off-switch that temporarily extinguished behavior upon achieving reward (Panksepp 1986, 1999, p. 243). Thus the catecholamine system (i.e. serotonin, dopamine, and norepinephrine) and its pathways came to be identified as the system of drives, and the endorphins and enkephalins and their pathways came to be regarded as the euphorigens of the brain reward system (Belluzi and Stein 1977; Stein 1978; Wise and Bozarth 1982). Pfaff (1999) went on to demonstrate that the drive system is actually a three-tiered set of interlocking electrophysiological, hormonal, and genetic mechanisms.

> During most of the twentieth century, neurobiologists appealed to the complexity of synaptic relationships and electrophysiological mechanisms to explain integrative properties of the nervous system, but we could not explain molar behaviors. Superimposed on the anatomical-physiological levels of explanation was a second level of explanation based on the neurochemistry of neurotransmitters and neuropeptides…. Now, based on the findings… in the hormonal control of reproductive behavior, we proposed a new level of neuronal integration superimposed on the first two. Interactions among transcription factors around well-recognized DNA response elements can embody the combinatorial logic required for the proper governance of instinctive behaviors. (Pfaff 1999, 139)

These achievements in the neurobiological theory of reinforcement helped to rescue drive theory from disrespect. Under Hullian, or drive reduction, behaviorism, drives were individuated and multiplied in an irresponsible manner that is currently reminiscent of meme theory.[20] Instead of a few primary drives and their conditioned derivatives, it was possible to invent drives for virtually everything. Pfaff has summarized the body of research on drive to show that this nested hierarchy of neural, hormonal, and genetic systems provides a mechanistic account of drive systems dedicated to sexual behavior:

> The principle established here is that a discoverable neural circuit underlies an integral mammalian behavior. In this circuit, natural biochemical (hormones) working in specific parts of the brain cause change in behavioral disposition. Reproductive behavior is

---

[20] E. G. Boring noted a similar difficulty with the reflex concept. Once drives or reflexes were decoupled from physiology a potentially infinite number of either could be hypothesized: "Not only may you have sugar-reflex as distinguished from a salt-reflex but you have a stuffed-olive reflex as different from an anchovy-reflex" (Skinner 1979, p. 145). See Walter (2007) where I discuss the manner in which ideas about everything can be turned into memes.

determined by estrogen and progestin effects emanating from medial hypothalamic neurons interacting with specific somatosensory signals. (Pfaff 1999, 193)

Thus, on the current neurobiological model, drives can be seen to be finite and specific hormonal and electro-chemical materialistic brain events produced by finite specific physical brain structures that activate genetic facilitators.

Only those who would endorse a Betty Crocker type theory of emotional consciousness would attempt to characterize these states as non-physical epiphenomena. This is easy enough to do if you buy into one of the forms of mind–body dualism. Rolls, for example, states his commitment to the dualist doctrine:

My view is that the relationship between mental events and neurophysiological events is similar (apart from the problem of consciousness) to the relationship between the program running in a computer and hardware of the computer. (1999, 264 note)

The appeal of the software metaphor is that it seems to provide a basis for flexibility in responses where outcomes are contingent on variation in circumstances. The program builds a bridge between the hard-wired reinforcement mechanisms and the contingent behavioral 'strategy' that is directed toward an adaptive outcome.

We have reached a quite specific view about how brains are designed around reward and punishment evaluation systems, because this is the way that genes can build a complex system that will produce appropriate but flexible behaviour to increase their fitness.... The way natural selection does this is to build us with reward and punishment systems that will direct our behaviour towards goals in such a way that survival and in particular fitness are achieved. By specifying *goals* [my emphasis], rather than particular behavioural patterns of responses, genes leave much more open the possible behavioual strategies that might be required to increase their fitness. (Rolls 1999, 284)

This is essentially the same explanatory picture that Dawkins provided in *The Selfish Gene* (1976), except that the reward mechanism intervenes between the gene and the behavioral strategy. The reward mechanism specifies the target, while the strategic program supplies the means. The strategic program appears to do this on the basis of what we expect from software programs: sentential attitudes and logical relations between said states. The concept of an *evolutionary stable strategy (ESS)* plays a crucial role in this formulation.

First, let us characterize the ESS the way it was originally formulated in population genetics. The ESS refers to examples of balanced polymorphisms where more than one variant is established in a population. A prime example of this is the balanced heterozygote that keeps sickle cell anemia in the population even though it has deadly effects. The situation that maintains the double recessive sickle cell is that the double dominant allele constructs blood cells that are susceptible to malaria, also a deleterious condition. Why do both these inferior alleles continue to exist in the population? The answer is that the heterozygote produces a blood cell that is impervious to malaria but also does not sickle. However, in order to get the healthy heterozygote, proportions of each inferior allele will have to be maintained in the population (Ridley 1985).

This type of cost-benefit trade off modeling was used to develop computer simulations of cooperation and competition in games theoretic treatments of behavior within a population (Maynard-Smith 1982; Ridley 1996). One paradigmatic case was provided by a simulation that established a stable ratio of 'hawks' and 'doves' within a population. On the first models, one is either a hawk that always fights, or a dove that never fights. The costs of fighting lead to a decline in the population of hawks and an increase in the population of non-fighting doves. Eventually the ratio of hawks to doves hits equilibrium. The computer simulations in these models track the outcomes of social interactions on the premise that the two alternative behavioral strategies can be linked to two alternative genes.

What made the software program of mind theory intriguing was the introduction of contingent behavioral strategies. That is, sometimes it might pay to act like a dove and other times it might pay to act like a hawk given the different circumstances that prevail in a given situation for a given social interaction. In conditional circumstances, 'strategies' were employed by actors to calculate the fitness payoffs for one or more lines of conduct. Thus, the computer simulations of population biology came to be turned into the psychological decision-making programs of the evolutionary psychologists. Since people are largely unconscious of the many factors that would have to be taken into account in making these decisions, the processes are either modeled on 'as if' intentionality—with recognition that the hypothetical processes are merely heuristic, or they are postulated to be serious and real but non-conscious mental processes. Let me illustrate with an example drawn from Rolls (1999).

Rolls begins his analysis by discussing some of the basic brain structures implicated in the reward circuit for sexual behavior in both laboratory rats and macaques. The medial preoptic area appears to be important for both males and females, although the dorsomedial hypothalamus plays a controlling role in sexual arousal and performance for the male, while sexual arousal and performance in the female is controlled by sites in the ventromedial hypothalamus. The dorsomedial area is active in the male during copulation, while the ventromedial area is active in females during presentation. Lesions to the dorsomedial area terminate mounting, intromission and ejaculation in the male. The administration of testosterone in this area restores sexual appetitive behavior. Rolls describes the connections of the preoptic area to other regions of the brain such as the orbitofrontal cortex and the amygdala. The former area is involved in face recognition and the latter area is involved in association learning connected to the sexual activity. Rolls concludes that the preoptic area is the site specifically involved in generating the rewarding effects of primary sexual rewards (Rolls 1999, pp. 221–222).

What we have above is a description of some part, at least, of the proximate mechanism for sexual reward. At this point, Rolls turns his attention to the sociobiology of the reward mechanisms involved in sexual behavior. This, unfortunately, is where propositional-attitude psychology makes its appearance in his argument. He discusses sperm competition and how this may affect male and female sexual behavior based on the research of Baker and Bellis (1995). It is clear that females may be able to determine which males impregnate them by the timing and sequencing of sexual activities with different males. It is also clear that males may be able to bias their chances of impregnating the female by a similar set of timed and sequenced events. What is not clear is whether these 'as if' decision processes engaged in by all the involved parties are 'as if' or whether they represent really existing decision processes. Rolls claims to endorse the 'as if' alternative (1999, p. 219), but, if that is the case the underlying processes by which these outcomes are achieved remain mysterious. In fact, once Rolls moves from discussing the nuts and bolts of the sexual reward circuit, which he does in strictly mechanistic terms, to the 'strategies' that use these mechanisms, his explanatory language shifts to that of rules and representations (Rolls 1999, p. 236).

Rolls' second example draws the commitment to rules and representations out more clearly. His first step is to cite genetic evidence in regard to the distribution of two alleles that influence infidelity in males. The gene affects the D4 dopamine receptor. Dopamine is one of the neurotransmitters involved in sexual reward. One allele comes in a long form, and men that have this version are more promiscuous than the men who inherit the short form (Rolls 1999, p. 231). Another gene that influences production of serotonin comes in two forms and the men who inherit the low 5HT producing version, which is associated with high anxiety, tend to engage in more sexual activity. Both examples show how a gene could influence the construction of some part of the brain's reward system and lead to differences in pattern of sexual behavior. These alleles could be distributed in a population such that their distribution forms an ESS. Differences in sexual fidelity linked to these genes could be an outcome of this 'strategy'. This is a simple mechanistic model.

Rolls, however, wishes to explore a hypothetical example that depends on conditional decision-making. He compares the 'strategies' of (1) faithful and (2) unfaithful males and those for (3) faithful and (4) unfaithful females. He proposes a number of pros and cons that individuals might consider in deciding whether to be faithful or unfaithful. There are a lot of factors to consider: Potential loss of resources if the infidelity is found out for the male perpetrator via divorce. Better genes than those provided by the husband if the adultery by the female is not found out. A number of the factors seem to lack a psychological dimension, however. Being born beautiful or handsome seems to dictate whether one is flirtatious or promiscuous more than anything else. Much of the decision-making seems to depend on non-strategic factors. However, Rolls argues that a man disposed to infidelity might be expected to be motivated to produce an orgasm in his partner and thus improve his chances in sperm competition against competitors whose sperm he needs to displace (Rolls 1999, p. 235). Rolls argues that an unfaithful female might be expected to experience reduction in desire for a particular partner, thus facilitating the search for new and better genes (Rolls 1999, p. 234). Rolls presents some of these 'strategies' as if the subjects are making decisions on the criteria that adduces to explain their behavior—even though those subjects cannot be said to be consciously intending to pursue these goals for those reasons.

If these examples are examples of 'as if' intentionality, then what brain processes are really going on? If, on the other hand, these examples exemplify real decision-making processes albeit at an unconscious level, what could be possibly cited as evidence to that effect? The latter alternative implies that there is a non-physical software program running on the brain's wetware, and there is no evidence that a software analogue for mental events running on top of the brain can be justified. Moreover, if there are mental events in a non-physical sense, then there can be no question of these events evolving by means of natural selection since they are not material entities governable by natural law. If, on the other hand, the examples are only examples of 'as if' intentionality, the purported processes do not really exist. On either scenario we have a vacuum which requires an explanation. Or do we? Perhaps we are in the same situation we were in with regard to Deacon's and Tomasello's criticisms of Chomsky.[21] The apparent sentential/computational complexity is an artifact of the complexity of social interactions outside the brain. The simplifying function in the case of emotional behavior is to be found in the reductive mechanisms of the evolved reward and drive circuits of the brain. There is no need to reify the features of our extended social interactions, thus turning them into some kind of Darwinian spaghetti code, projecting them back into the 'mind' as a system of rules and representations, albeit directed toward fitness goals.

Why Proximate Mechanisms Matter

The critical criterion of a valid evolutionary explanation of behavior is that it has to distinguish a just-so hypothesis from a justified explanation. Sociobiology failed to do this because its practitioners stipulated that the concept of adaptation guaranteed that all evolutionary outcomes were adaptive. In their criticism of sociobiology, evolutionary psychologists were right to bring the ancestral history of proximate mechanisms back into focus. Adaptations must be linked to the circumstances of their ancestral history; yet, there is no guarantee that yesterday's adaptation is not today's evolutionary dead end. 'As if' intentionality is no better than sociobiological fitness-maximizing stipulations that guarantee reproductive success because they provide the same type of pseudo-guarantees that behavior follows an adaptive program whether that program is currently adaptive or not. The model is infinitely malleable and can pretend to explain any and all outcomes. Hence, the explanatory problem is clear: why substitute population genetics in the form of IPM psychology for a theory of proximate mechanisms?

In addition, information processing models simply are not a good candidate for a theory of evolutionary proximate mechanisms because of their ontological commitment to a dualism of mind over brain (Churchland 1986; Uttal 2004). There is no software analog that runs on top of the wetware of the brain. Despite the fact that our species engages in linguistic behavior, the brain accomplishes its work, both linguistic and non-linguistic, via a physical, non-sentential,

---

[21] See pp. 16–17 in Chap. 1.

electrochemical system, possibly in accordance with mathematical descriptions supplied by tensor network analysis as suggested by neurobiological materialists such as Churchland (1989, 2007) or Grossberg (1988).

Folk Psychology Redux

I think we are now ready to squash the idea that folk psychology can provide an adequate explanation of proximate motivation. Let us revisit Searle's thesis that intentional states explain our behavior. Recall that both Millikan (1993) and Sterelny (2003) line up behind folk psychology here. Are belief-like intentional states what the evolutionary psychologist needs in order to cover proximate explanation? According to Kitcher (1985) and Vayda (1995, 2008) folk psychology is the right kind of psychology and it is evolutionary psychology's task to prove it so. I would like to dispel that endorsement by demonstrating what the limits are of an action-focused or intentionality based explanation. Let me illustrate using an example that will demonstrate the central role that proximate mechanisms conceived as mechanistic brain processes play in the explanation of adaptive behavior.

Sober (1985) has argued that the *idea* of incest avoidance functions as the proximate mechanism that functions in the human case the same way that a non-sentential neurological mechanism functions in a non-verbal animal (Sober 1985, p. 188). However, the terms in the two examples are not equivalent. To state one's intention is not to explain why one has that intention. To desire something is not to explain why one desires that thing. I may state that I do not wish to marry my sister and that I do not wish to marry my sister because I have no desire for her, but such beliefs and desires do not themselves constitute a proximate account of inbreeding avoidance. If humans demonstrate a desire to avoid incest, this may involve a specific neurological mechanism dedicated to inbreeding avoidance.

For instance, Demarest (1983) has suggested that the mechanism that results in inbreeding avoidance is a neural circuit that involves the amygdala and the anterior hypothalamus. Demarest, however, does not believe that this circuit can be called an inbreeding avoidance mechanism. It could only be ascribed that status if it could be shown that the neural circuit that is implicated in the behavior exists because it produced a non-inbreeding outcome. According to Demarest, the non-occurrence of inbreeding is an unintended consequence of some other evolved function for that circuit. In this case, Demarest believes the circuit is simply habituation. Familiarity breeds contempt, not because of the maladaptive consequences of inbreeding but out of sheer boredom with the familiar. My own research on the topic of inbreeding avoidance suggests that male and female humans differ in their inbreeding avoidance dispositions.[22] Males show little

---

[22] Walter and Buyske (2003) update the models that were presented in Walter (1997).

negative impact of close childhood proximity on adult sexual desire for female
cousins whom they saw everyday, but females show an intense aversion to mate
with male cousins whom they saw everyday in childhood. Chimpanzees show this
same pattern (Pusey 1980). The higher cost of inbreeding to females in ancestral
environments may explain why there is a sex difference in inbreeding avoidance.
Such questions belong to the level of the 'design' stance. To justify the attribution
of an evolved proximate mechanism dedicated to inbreeding avoidance would
require that the nature of the neurological mechanism be specified. Perhaps males,
being less prone to inbreeding depression in ancestral environments, have a
Demarest type circuit. Familiarity breeds contempt but only out of boredom.
Females, on the other hand, may have a *bona fide* inbreeding avoidance mechanism because of the high cost of inbreeding in their environments of evolutionary
adaptation. The fact that female chimps only begin to avoid their male playmates
and fathers and emigrate from their natal groups when they begin to cycle
sexually, suggests that ascription of an inbreeding avoidance mechanism may be
justified. We would need to know more about the neuro-hormones and other brain
circuits that are implicated in order to assert this with confidence (Steklis and
Walter 1991; Walter 1997, 2000; Walter and Buyske 2003).

The main point is this: we need to know sufficient detail about the proximate
brain mechanisms involved in producing a behavior in order to claim that it is an
adaptive mechanism of one sort or another. Reasoning backwards from the
non-occurrence of incest alone opens us up to making spurious ascriptions of
function where none may exist or some other function may be at work. As Quine
(1989) counsels, we do not need to specify the full neurological story, but we do
need to specify enough detail to warrant an ascription of function. Folk
psychology, whether in the form of Dennett's 'as if' intentionality or Searle's
intrinsic intentionality, is not equipped to perform such a role in the explanation
of behavior.

While the folk-psychological idiom may not be eliminable in ordinary everyday, non-scientific, talk, eliminative materialists such as the Churchlands, or
behaviorists such as Quine, are surely right to eliminate it from causal-mechanist
scientific accounts of behavior. The mechanist perspective is an attempt to
resuscitate the neuro-ethological commitment to middle range causal mechanist
accounts of behavior where proximate mechanisms are specified in enough neurological detail to justify the attribution of evolutionary adaptation where
warranted. I think this *mechanist stance* will provide a more satisfactory paradigm
for the explanation of behavior than the propositional-attitude alternative.

The 'Psychologic Gambit' Versus the Mechanist Stance

Some readers may be inclined to object that I have overstated the case that evolutionary psychologists have systematically neglected the domain of motivational
endowments. They will point to the large body of research on Fluctuating

Asymmetry (FA) that is summarized by Thornhill and Gangestad (2008). They may also point to a small industry in research devoted to 2D:4D ratio (Robinson and Manning 2000; Puts et al. 2004). To be sure, these burgeoning concerns show a promising trend in the discipline. But the fact of the matter is that the proportion of research that includes a consideration of hormonal factors and other neural based research is small compared to the large proportion that is predicated on what Puts (2010) calls the 'psychologic gambit', and which I have been referring to as 'propositional-attitudes'.

Towards the beginning of this chapter, I commented on the fact that there is no treatment of the emotions in *The Handbook of Evolutionary Psychology* (Buss 2005). This despite the fact that the edited volume covers key issues pertaining to mating, parenting, aggression, social exchange, and to other factors central to survival and differential reproduction. Despite the omission of emotions, there is some attention devoted to hormonal mechanisms,[23] but even so, the discussion of neural machinery is kept to a minimum. How can this be so? Puts (2010) attributes the lack of neural focus to the 'psychologic gambit' which refers to the belief that the neural details don't really matter:

> This is the implicit assumption that neurophysiological underpinnings can largely be ignored when testing evolutionary hypotheses about behavior and psychology. After all, if one is interested in the evolutionary functions of behavioral and psychological patterns, does it matter whether these patterns depend upon the nucleus accumbens or the baso-lateral amygdala, for example, or upon vasopressin or oxytocin, when selection only 'sees' the behavior? (306)

The principal idea behind the 'psychologic gambit' is that *function* belies mechanism. If the mechanism merely serves the function then the mechanism can be left as a black box. Thus, although Tooby and Cosmides call for the inclusion of neural, developmental and genetic factors in evolutionary psychological explanation in their opening contribution of the state of the art *Handbook* (Tooby and Cosmides 2005, p. 6), the call appears to be optional. If, however, function (i.e. adaptation) is as onerous an attribution as Williams (1966) claims it is, the devil frequently will lie in the developmental details. This devil has most recently emerged in the challenge that evolutionary developmental systems, otherwise known as 'evo-devo', has issued to its psychological allele.

Evo-Devo Versus Evolutionary Psychology?

The theory of Evolutionary Developmental Systems, christened 'evo-devo' by some of its practitioners, such as Griffiths and Gray (1994), proposes an explicit challenge to the 'adaptationist' perspective offered by evolutionary psychology. What is the issue? After all, we are all Darwinians here, are we not? What could

---

[23] For example, Gangestad et. al. (2005) and Flinn et al. (2005).

possibly oppose developmental biologists to evolutionary psychologists? Briefly, the 'evo-devo' camp accuses the evolutionary psychology camp of accepting a pre-formationist view of biological development, which implies that biological development consists of the unfolding of a pre-determined *program* that does not recognize the contingent and mechanistic nature of developmental processes. Lickliter (2008) thus argues that evolutionary psychology must be recast with appropriate attention to contingent developmental processes. Certainly, Fodor's version of cognitive modularity (Fodor 1983), which does not countenance any form of empirical contingency seems ripe for criticism from Evolutionary Developmental Systems Theory. Do evolutionary psychologists buy into the same Platonic paradigm—and if they do so, why? The Platonic commitment is easy to find, and Lickliter and Honeycutt (2003), who are committed 'evo-devoists', attempt to get to the crux of the problem—and they locate it in a statement made directly by Tooby and Cosmides themselves:

> [T]he individual organism, fixed at conception with a genetic endowment regulating its developmental *programs* [my emphasis], encounters its specific ontogenetic environment, which it processes as a set of specific inputs to these developmental programs. In other words, the organism blindly executes the program it inherits, and the ontogenetic conditions it encounters serve as parametric inputs to these programs. (Tooby and Cosmides 1990, 388 quoted in Lickliter and Honeycutt, 2003, 360–361)

For 'evo-devo' practitioners, the problem resides in the pre-set nature of the developmental program. One of its consequences is that contingent details of evolutionary development are disregarded as mere glitches in an otherwise smooth running program. According to Buss (Buss and Reeve 2003) and Tooby and Cosmides (1990) the role of development serves primarily to guarantee that the ontogenetic outcome achieves a position within its predestined normal range of variation. Genuine evolutionary contingency thus appears to apply only to selection pressure at the level of phylogeny.

What is really at stake here? Both the evolutionary psychologist and the evolutionary developmental systems biologist are committed to *flexibility* in evolution. Both oppose rigid pre-determined solutions to complex adaptive problems. For the 'evo-devoist', however, the flexibility is located in the contingent interactions between genetic and epigenetic interactions during development, while the evolutionary psychologist locates the source of adaptive flexibility in 'decision-rules' that appear at the culmination of the developmental process (Buss and Reeve 2003, p. 850). Hence, the crux of the problem with the developmental program concept resides in the program concept itself. Developmental plasticity, which is mechanistic and contingent, is sacrificed in order to preserve the behavioral flexibility of the fully developed cognitive program. The problem with the *program* concept, however, is not only that genetic and epigenetic interactions cannot be modeled in terms of a program, but also that the activities of the mature brain itself do not operate in such terms. As Patricia and Paul Churchland constantly remind us, there is nothing in the dynamics of neural networks that can justify the deployment of the software program concept in the explanation of cognitive representation. Are there evolutionary psychologists who manage to avoid this mistake?

The Physiologic Gambit in Evolutionary Psychology

The pivotal role of specifying neural mechanism is made clear in that research within evolutionary psychology that actually does appeal to it. I am thinking here of research on 2D:4D ratio and that on Fluctuating Asymmetry. First, with respect to 2D:4D finger length ratio, Robinson and Manning (2000) present research that supports the hypothesis that prenatal levels of testosterone are associated with homosexuality and low 2D:4D ratios. A masterful review of the research on 2D:4D ratio by Puts et al. (2004) enabled a sharpening of focus on the mechanism and what is and what is not entailed in its functional scope. Their review reveals that much of the research provides mixed or contradictory results. Two of three studies with female subjects show the opposite direction of effect than that which was predicted for a number of sex-linked traits, while one of two studies with male subjects show a similar reversal in direction from the predicted one. Puts et al. (2004) conclude that these mixed results show that only sexual orientation itself, and not any other sex-linked traits, are reliably associated with low 2D:4D finger ratio. They further conclude that the determination of sexual orientation is linked to the timing of developmental events specifically involving testosterone. Thus, even though the authors of the review collected no hormonal data for the study, they were able to use pre-established hormonal data in conjunction with a winnowing of irrelevant behavioral traits to find the relevant hormone-behavior link. One has to ask where the theory of 2D:4D would stand if the hormonal mechanism that underlies the observed behavioral preference was not known?[24]

A second field of research within evolutionary psychology that utilizes hormonal data in its theory building involves Fluctuating Asymmetry (FA). FA has provided a major industry for the discipline. Much of the research in this field has been conducted by Steve Gangestad and associates and is summarized in a book that he co-authored with Randy Thornhill entitled, *The Evolutionary Biology of Human Female Sexuality* (Thornhill and Gangestad 2008).[25] The principal thesis

---

[24] I have to admit that I am perplexed by the claim that low 2D:4D ratio is reputed to be associated with homosexuality in general. I would expect and hence predict that the ratio depends for each sex on which gender role the individual prefers. In other words, I would expect that males and females who prefer to take the male role in a sexual encounter would exhibit low 2D:4D ratio (as some female homosexual subjects seem to exhibit), and that females and males who prefer to take the female role will exhibit a contrary high 2D:4D ratio. One could take a sample of males and females and partition it into homosexual subcomponents, then partition that into male vs. female role preference and then associate those partitions with 2D:4D finger ratio. A hierarchical log-linear or logit procedure might be useful in finding such relationships.

[25] See Gangestad et al. (2005) for the short version.

of FA is that female sexual preferences vary over the course of their sexual cycle such that during the peri-ovulatory phase they evince a preference for men who possess symmetrical faces, robust facial features, and odors that emanate from men with higher testosterone levels. These are men who portend signals that advertise that they possess high quality, and therefore desirable, genes. When females are not ovulating, they are hypothesized to not prefer such men; instead, they prefer both men and women who are kin because the latter can be relied upon to provide needed long-term support. A low ranking male who lacks who lacks the desired ensemble of masculine traits might be able to mate successfully on the basis of being a good provider. FA theorists argue that Fluctuating Asymmetry disposes females to prefer and possibly seek extra-pair copulations (EPC) with men at the high end of the masculine spectrum. That way they can get good genes but not sacrifice long-term support from men at the lower end of the spectrum whom they married for practical reasons of enduring support (Gangestad and Thornhill 1997; Gangestad et al. 2010). The primary experimental procedure is to correlate day in the ovulation cycle with a shift in preference for symmetry and masculinity. One important set of cues is hypothesized to be olfactory. Hence Garver-Apgar et al. (2008) obtained a correlation between high levels of estrogen during ovulation and a preference for the scent of males who have been judged to be highly symmetrical.[26]

Critics of FA such as Dixson (2009) raise questions about the quality of the research on Fluctuating Symmetry that must be taken seriously. Rantala et al. (2006) found that only cortisol was associated with female preferences in regard to male odor stimuli, and that this preference was present throughout the entire menstrual cycle and not just in one phase of it. They also found a distinct aversion in females to strong masculine odors that was also present throughout the menstrual cycle.[27] This research would seem to send the FA school packing. Dixson (2009), who is a primatologist, and a researcher committed to a multidisciplinary approach that includes endocrinology, argues that the 'ovulatory shift hypothesis' advanced by Gangestad and Thornhill (1997) is poorly supported. He argues that it does not make sense for females to jeopardize long-term reproductive investments for the sake of a quick but very risky infusion of 'good genes'. He argues that the hormonal data do not support the theory and neither does the argument regarding the ancestral environments that selected the hominid female mating 'strategy'.[28]

---

[26] It should be noted that the researchers did not collect hormone data from the subjects in the study but relied upon previously established schedules of ovulation cycle phase and hormone level.

[27] It should also be noted that Rantala et al. did collect hormonal data from their subjects.

[28] I refer the reader to two of my own papers that offer further reflections on the issue: Walter (2000, 2002).

Conclusion: The Psychologic Gambit in Decline or Reductionism at Last!

Whether you are a supporter or a critic of FA theory, one positive development in regard to theory construction in evolutionary psychology seems possible: an increased commitment to mechanism as opposed to intentionality and other forms of propositional-attitude. This turn to mechanism, however, requires the rehabilitation of one concept that has been claimed by the propositional-attitude school as its own, and that is the concept of *preference*. On the first page of the first part of this essay, I summarized Jerry Fodor's *Language-of-Thought* model. 'Preferences' were there accorded a key role in the scheme. On that model, once you recognize your preferences, they can be realized by an assortment of belief-desire schemes. For example, someone who is committed to the LOT model would take an ovulatory shift in preferences and re-contextualize it in terms of a temporal shift in beliefs and desires. What shifts there are, if any, reveal that the subjects who experience these phenomena fail to develop the sort of preferred description that is required for intentionality as advanced by Searle (1984). One would have to be deeply committed to the notion of an unconscious information processing program to defend the LOT model. The decoupled representation model proposed by Sterelny (2003) also fails in its attempt to subjugate motivational endowments to representations because it is clear that learning is on a leash that is controlled by the brain reward mechanisms. We need not postulate some form of internal decision-making homunculus that picks and chooses which preferences on which to act and how. My suggestion is to resist conceptualizing preference as a sub-species of propositional-attitude. Instead, we should assimilate preferences to the notion of behavioral disposition. As a consequence, preferences and dispositions can be conceptualized as motivational endowments which are better conceived in terms of brain mechanisms than as sentences in a programming language. This requires a vigilant commitment to the mechanist stance.[29] Thus, until the 'Psychologic Gambit' is rejected, I shall decline to call myself an evolutionary psychologist.

---

[29] I say vigilant because even when there is a strong principled commitment to the mechanist stance, there is still a tendency to wax poetic with propositional-attitudes. For example, Peter Ellison, who is one of the editors of the recently published and highly monumental mechanist tome, *Endocrinology of Social Relationships* (Ellison and Gray 2009), can be found to characterize hormones as carrying *information* about the state of an organism such that adaptive *allocation* of reproductive effort can be made in response to those states. Is there someone present who is making decisions about how to allocate resources? Or is evolution itself capable of such propositional-attitudes? This problem, of course, arises when trying to connect the proximate part of the explanation to the ultimate part. However, the solution is not to turn the environment of selection into a purposive selector that executes decision-rules. Such a move cannot but turn evolution into a process akin to the dialectical unfolding of Hegel's *Geist*.

Chapter 3
Postscript: The Virtues of Weak Modularity

'Modularity' refers to the encapsulation of information within specific systems such that these systems are isolated from one another. A less Platonic way of phrasing this would be to claim that biases exist in neuro-cognitive response systems that are geared to specific classes of stimuli. Fodor (1983) advanced the view that perceptual and sensory input systems are strongly modular but that cognitive systems are not. The latter are domain-general rather than domain-specific. Against the strong-modularity[30] perspective offered by Fodor, there are theories of cognitive modularity that hypothesize that central processes are also encapsulated, but not impervious to training. These come from the laboratories of neuropsychologists such as Gazzaniga (1985), Allen (1983), and Hellige (1990), who have researched hemispheric lateralization in split-brain patients and from clinical work with brain-damaged patients (see, e.g., Damasio 1994). Gazzaniga for example, uses the term module to describe the independent functioning and localization of brain systems that underlie cognitive activities such as language, facial recognition, and emotions. For Gazzaniga, language is one module among others.

Cosmides and Tooby's ground-breaking research in the field of social cognition seemed to demonstrate that human subjects are much better at understanding the implications of a question that involves social cheating than they are at using logic to understand an abstract question with an identical logical structure (Cosmides and Tooby 1992, 1994). This led them to postulate a 'cheater-detection' module and to argue for its reproductive utility in ancestral environments. Cosmides and Tooby claim that the mind is like a 'Swiss Army-knife,' with as many domain-specific modules as there were adaptive problems in ancestral environments. They argue that there is no domain-general cognition because there are no domain-general problems to be solved. Note the almost cavalier way in which modules are

---

[30] To run ahead a bit, the reader will shortly learn that 'strong' modularity and 'massive' modularity are not synonymous. 'Massive' modularity refers to the extension of modularity to higher brain function.

A. Walter, *Evolutionary Psychology and the Propositional-attitudes*, 51
SpringerBriefs in Philosophy, DOI: 10.1007/978-94-007-2969-8_3,
© The Author(s) 2012

imputed for various functions: "incest avoidance, social exchange, aggressive threat, parenting, mate choice, disease avoidance, food aversions, predator avoidance, habitat selection, and so on" (Cosmides and Tooby 1994, p. 105). The "and so on" is quite clearly an invitation to go forth and multiply. Thus, an important question to address is how exactly is such a diverse group of mechanisms organized? For example, are mate choice and incest avoidance modules sub-domain-specific-sub-modules of one mate selection? Or are they independent? Confusion notwithstanding, one is obligated to give the evolutionary psychologists enough time to work out the finer details of their position. Nevertheless, it is with considerable optimism that Cosmides and Tooby proclaim that evolutionary and cognitive psychologists are like two teams of explorers working their way up an uncharted mountain from opposite sides, surprised to face each other when they converge at the top (Cosmides and Tooby 1994, p. 85). They will all be shocked upon arriving at the summit only to find Jerry Fodor and his granny already there in full headdress, warpaint, and loincloth with spears pointed directly at their reproductive organs.

In *The Mind Doesn't Work that Way: The Scope and Limits of Computational Psychology* (2000), Fodor specifically targets the 'massive modularity' thesis.[31] He issues this challenge on the basis of 'abduction'. 'Abduction' refers to the process of finding the best evidence to justify a belief no matter where it is located. A massively modular mind would have limited ability to perform various feats of abduction. Some mental organ must exist that integrates all the data. Only in this way would it be possible to successfully generate the appropriate abduction.

Despite the best efforts of cognitive scientists over the past 50 years, Fodor finds the issue of the structure of the mind to be an unsolved mystery (2000, p. 99). He specifically challenges the validity of Cosmides and Tooby's interpretation of human performance on the Wason task and criticizes both the domain-specific hypothesis and evolutionary psychology in general. Fodor's critique is predicated on his view that cognition is independent of evolutionary and neurophysiological processes. Fodor argues that the qualitative saltations in human cognitive evolution could not be accounted for by Darwinian gradualist monotonic changes in the genes that construct our brains. In a review of an earlier work of Fodor's, *The Elm and the Expert* (1994), Steven Pinker registered some surprise that Fodor expresses explicit hostility to the role that Darwinism portends to play in explaining the evolution of the structure of the mind. Pinker speculates that this hostility is predicated on a "contemporary allergy to Darwin in the humanities and the cognitive sciences." (Pinker 1995, p. 205) In actuality, Fodor's hostility to neurophysiology and evolution should come as no surprise, since he openly admits that he is a Cartesian dualist. Fodor believes that ideas or mental representations are essences that exist on a different level from whatever physical substance they happen to be momentarily housed in. For Fodor, cognitive processes do not consist

---

[31] I believe the term 'massive' was first proposed by Fodor (2000) to apply specifically to the evolutionary psychologist's conception of domain-specificity.

in whatever the brain is doing but rather in the processes that comprise the essential properties of 'mentalese'.

However, one has to wonder why, since Fodor's view of modularity has always been conditioned by his commitment to transcendental nativism, he should bring forth arguments against modularity in *The Mind Doesn't Work That Way*. The short answer is that evolutionary psychologists such as Pinker (1997) see the evolution of modular cognitive systems as being explainable on Darwinian grounds, and Fodor simply does not want Darwinian nativism to displace transcendental nativism (see Fodor 1998, pp. 203–214). Since Pinker sets the research agenda for the evolutionary psychology paradigm as fleshing out the modular organization of the human mind, Fodor must counter with cognitive processes that have no Darwinian explanation. Hence his attack in The *Mind Doesn't Work That Way* on what he calls the 'massive modularity thesis.' Fodor describes the stakes as follows: "if the cognitive mind is massively modular—if, that is, the mind is exhaustively a collection of modules—then psychological Darwinism must be pretty generally true of it" (2000, p. 6). Although he is willing to admit that some parts of the mind are modular, he maintains that mentalese in general is abductive. Much then hangs on the prospect of finding general, unencapsulated cognitive processes. What would constitute an example of one? An example given by Fodor is that of balancing one's checkbook (1998, p. 155). Showing that abduction is central to cognition is what Fodor believes makes his case against cognitive, or massive, modularity stand or fall.

Let me give a personal example to illustrate the contrary thesis that abduction is weak or nonexistent. The author of this essay remembers when he was once taking a statistics exam and floundering over a probability question. He remembers producing a number of unhelpful sentences; finally he sat defeated in front of the story-problem. To his surprise, his right hand began to move, and he watched it produce what turned out to be the correct answer. I believe this example illustrates the counter-thesis that mathematical reason is nonverbal, at least somewhat encapsulated, and hence abduction plays a minimal role, if it even exists at all. If anything, a good story-problem demonstrates that both verbal and mathematical 'modules' can be simultaneously active, working on different aspects of a given problem. In my case, attempting to employ the wrong module on the wrong task resulted in temporary confusion and failure.

Research that supports the hypothesis that there is a distinct mathematics module has been conducted by Cantlon et al. (2009). Using MRI techniques, their research reveals that children aged 6–7 activate the same areas of the brain as that are active when adults perform numerical notations over both symbolic and non-symbolic calculations. Children in this age group also employ regions in the inferior frontal cortex more than do adults. This suggests the existence of a developmental shift in module organization. In 2011, a special issue of the journal *Developmental Neuropsychology* was devoted to the development of mathematical competence in ontogeny (see Ansari 2011). I believe this research can be interpreted to support the view that a mathematics module exists, but that the module is developmentally labile as argued by developmental theorists (e.g. Marcus 2004 and Karmiloff-Smith 1992). Is there a phylogenetic side to this story?

Portrait of the Mind–Brain as an Evolved Epigenetic System

Steven Mithen's book, *The Prehistory of the Mind* (1996), combines an attempt to flesh out some of the contours and details of the 'environment of evolutionary adaptation' (EEA) that evolutionary psychologists are fond of referencing, and it also weighs in on the modularity issue. Fodor (1998, pp. 153–160) while reviewing Stephen Mithen's book also directed criticism at the modularity views advanced by Byrne and Whiten (1988) and Humphrey (1992). The latter two treatments postulate fewer, but nonetheless still domain-specific, forms of cognition. These are the *technical*, which covers intuitive physics, the *natural historical*, which covers intuitive biology, the *social*, which covers intuitive psychology, and *language*. Byrne and Whiten produced their landmark volume of papers in 1988 (which included an important foundational paper by Humphrey) that showed that modern chimpanzees had highly developed skills at social deception and manipulation compared to their undeveloped technical and natural historical skills. Mithen uses this example to provide the framework for understanding the evolution of the prehistoric mind. He also discusses the existence of what he refers to as general intelligence.

A crucial analytic tool deployed by Mithen is provided by Karmiloff-Smith's concept of cognitive fluidity. In *Beyond Modularity* (1992), Karmiloff-Smith argues that cognitive domains are not hard-wired but developmentally plastic. The infant starts out with highly modular cognitive systems, but these are replaced by integrative forms of cognition that are capable of representational redescription. This is made possible because of developmental plasticity in the original modules. Against this, Fodor argues that the developmental plasticity of the human mind during ontogeny may be less plastic and more maturational than recognized by Karmiloff-Smith (Fodor 1998, p. 129). Thus Fodor wants to maintain the nativist commitment against Karmiloff-Smith's empiricism, which holds that changes occur that are not preprogrammed. She is attempting to integrate Piagetian constructivism with Chomskian nativism. While Karmiloff-Smith admits that there might be generalized, non-modular cognitive processes, she is primarily concerned with demonstrating that cognitive development is not general but differs relative to each given module (Karmiloff-Smith 1994, p. 702).

Alternatively, according to Greenfield (1991), the 2-year-old child is not modular, but is operating on the basis of general intelligence that is eventually superceded by specific and diverse cognitive competencies. Cognitive fluidity involves all of these diverse competencies working seamlessly together. This appears to be identical to the way different modules work together in the unbisected intact brain according to Gazzaniga. Fodor would no doubt agree, but he would deny that the processes are modular if they are working together. The question then is whether diverse modules could work separately but simultaneously without a centralized integrator. Following Greenfield, Mithen argues that modern human ontogeny recapitulates hominid phylogeny. He argues that anatomically modern humans are capable of cognitive fluidity and that premodern humans were not. Anatomically modern humans demonstrated that they were capable of cognitive fluidity when they evolved the capacity to represent thinking from one domain in terms drawn from another. Modem humans became capable of analogy and metaphor, and we see evidence of this directly in the fossil record. Prior to that, they were incapable of it.

Mithen argues that our common ancestor, the Pliocene ur-ape with the *chimpanzee-like* mind, performs quite poorly in terms of domain-specific cognitive competencies. Mithen argues that chimpanzee tool use is too primitive to postulate any degree of technical intelligence. He suggests that only an elementary form of general intelligence (e.g., associative learning or trial and error) is necessary to account for their ability to pull branches off a twig in order to stick it in a log to pull out termites (Mithen 1996, pp. 74–78). He also argues that their natural history intelligence is only partly developed, because even though they do forage for food, they seem to possess only a limited level of ability to mentally map the location of these resources, and they show inflexibility in pursuing them. He also argues that they do not use a great deal of available visual information when they hunt prey. One might think that using a tool to fish for termites might actually be evidence that they integrate technical and natural historical knowledge, but Mithen is unwilling to accept this possibility (Mithen 1996, pp. 78–81).

On the other hand, as various contributors to the Byrne and Whiten volume on Machiavellian social intelligence have demonstrated, chimpanzees are experts at social deception and are cunning, ambitious, and deceitful. Therefore, chimps do possess a highly developed social intelligence (Byrne and Whitten 1988). Their relative inability to teach each other how to crack nuts effectively and the fact that they do not exploit material culture to advertise social status shows that their social intelligence is highly encapsulated from their general intelligence. It is important to note that chimpanzees already possess a high level of social intelligence, since Cosmides and Tooby (1994) argue that the evolution of hominid social intelligence occurred during the Pleistocene and was selected for on the basis of adaptive challenges faced by the demands of hunting and gathering on social cooperation. Fodor ridicules this assertion (2000, pp. 101–104). Instead, Fodor asserts that humans have a high level of social intelligence but that this is the result of a discontinuous qualitative "saltation" in human intelligence that cannot be explained by Darwinian evolution (Fodor 1998, pp. 165–167).

Fodor and Mithen (not to mention Pinker) hold that linguistic intelligence is uniquely human. Mithen claims that purported linguistic skills of chimpanzees are overrated and that they show an upper level of competence equivalent to that of a two-year-old human suggesting that they are accomplishing these feats solely on the basis of general intelligence. A challenge to this assumption comes from the work of Sue Savage-Rumbaugh on the linguistic competence of bonobos (Savage-Rumbaugh and Lewin 1994). If Savage-Rumbaugh is correct, bonobo linguistic ability is much more advanced than characterized by Mithen. Because bonobos in the wild, and presumably our common ancestor in the Pliocene did not use language, the implication is that linguistic intelligence is part of a more general form of intelligence that evolved in a nonlinguistic context. This would appear to be a blow to both Fodor's modularity thesis and to the domain-specific arguments advanced by evolutionary psychologists.

The modularity thesis, however, could be salvaged in the following way. Utilizing a developmental systems approach, Marcus (2004) has argued that cognitive systems are not evolved from scratch. That is, one system, say, language, may spin off from another non-linguistic system and hence both will share some of the same genetic catalysts. Marcus suggests that genes specific to language may be very few, and that most genes operating to construct linguistic systems also operate in the construction of other systems (2004, pp. 131–140). However, the sharing of genetic construction processes across different domains does not entail that the mind is a mono-morphic entity. The genetic differences, though numerically small compared to the aspects of development that are shared, make for a highly differentiated set of cognitive systems. Hauser, Chomsky, and Fitch (2002) hypothesize that cognitive systems involved in navigation (i.e. spatial intelligence), arithmetic, or social relations would be systems that could provide the basis for the recursive component in language. Tomasello (2008) argues more convincingly that linguistic recursion originated from a phylogenetically prior foundation of gestural communication. The question then becomes to what extent have various cognitive functions diverged from one another and why?

Hominid Evolution and Cognitive Fluidity

For the most part, Mithen's goal is to describe the phylogeny of modularity through a primate baseline analysis of cognitive generality. With the advent of the earliest members of the Homo lineage, which Mithen subsumes under *Homo habilis*, we see a qualitative increase in the modularization of cognition. These developments are tied to the unique challenges posed by the features of the Plio-Pleistocene adaptive landscape that facilitated the speciation event. While the first hominid toolkit, Omo industry, does not show any special cognitive specialization, the one that followed, Oldowan industry, does. Striking a flake from a core requires an understanding of fracture dynamics. Chimpanzees cannot do this. We believe that they have an extremely low level of technical intelligence because there is no evidence of patterning and the industry was static from start to finish (Mithen 1996, p. 98). With respect to natural history intelligence, the addition of meat protein to the diet via scavenging and hunting led to significantly increased skill levels. This is due to the demands of predicting and tracking the location of moving resources, the availability of which is also highly variable due to seasonal fluctuations in climate. Competition with predators over the same resources was also an important variable. Mithen here does a good job of steering through the acrimonious debate surrounding the evolution of hominid subsistence strategies (cf. Binford 1985; Blumenschine 1987; Isaac 1978; Potts 1988). It is possible that *Homo habilis* may have transported tools and meat to safe locations for processing (Potts 1988). They may also have carved out a specific niche in which to acquire animal protein at certain times of the year (Blumenschine 1987).

These factors have implications for the evolution of hominid social intelligence. In primates, as Aiello and Dunbar (1993) have shown, group size predicts brain size. As group size increases, one needs to be able to devote more neural capacity to remembering social relationships with other group members. Risk from predation, combined with the availability of animal protein in large, though relatively infrequent packages, would predict an increase in group size. Mithen argues, following Aiello and Dunbar (1993), the energy load on social grooming in these larger groups may explain the evolution of language, and that language may have originally functioned to ease group tensions. Analysis of the brain case of 2 million year-old KNM-1470 (see Falk 1983) suggests that *Homo habilis* had a developed Broca's area (an area of the brain which is implicated in speech production). Mithen seems to suggest that the earliest human linguistic intelligence is at base social intelligence. Despite the interrelated complex of forces that led to tool manufacture, the acquisition of animal protein, and the evolution of group size and communication, Mithen maintains that the cognitive faculties that support these activities remained isolated from one another (Mithen 1996, pp. 113–114). Tools show no evidence of any putatively social use that would extend beyond the domain of their narrowly constrained technical function of butchering carcasses. Mithen also maintains that the relationship between the intelligence involved in making the tool and that in acquiring the meat is also decoupled. Any connection there is can be accounted for by his theory of general intelligence.

Concerning the following phases of human evolution, Mithen makes the unusual move of combining all the putative hominid species within one classification that he calls "Early Humans" (Mithen 1996, pp. 116–117). This includes what most palaeoanthropologists describe differently as *Homo erectus, archaic Homo sapiens, and Homo sapiens neanderthalensis*. Those of us who follow Wolpoff s multiregional model of human variation (Wolpoff 1999; Wolpoff and Caspari 1997) will not have a problem with this, but others might be expected to take issue. Mithen's reasons for conflating the memberships of these various groups appear to derive less from Wolpoff than they do from his own scheme of cognitive organization. That is, he finds these "species" similar enough to each other, but distinct from either *Homo habilis* or anatomically modern humans in terms of their cognitive architecture to justify their being grouped together. That does not imply, however, that they are identical with respect to cognitive architecture. Early Humans, usually classified as *Homo erectus*, differ from earlier *Homo* specifically in terms of the degree of their technical intelligence as measured by their tool technology. The bifacial handaxe was a big improvement over the comparatively crude Oldowan core/flake technology. There was symmetry and design in the former that was lacking in the latter. This tool technology remained static, as did brain expansion, for approximately a million years, suggesting that a cognitive plateau had been achieved that was not exceeded until archaic *Homo sapiens* appeared between 400,000 and 500,000 years ago. At this time, several different kinds of intelligence received a big boost. Mithen claims that substantial brain expansion occurred that correlated with the indisputably more complex Lavallois tool kit. On the basis of this increased complexity in the tool industry, Mithen equates Early Human technical intelligence with that of modern humans. Moreover, based on the Neanderthal hunting skills that are evidenced by the fossil record, Mithen also attributes a natural history intelligence equal to that of anatomically modern humans. The social intelligence of Early Humans is unproblematic for Mithen because he already finds it present in earlier Homo and present even in the common ancestor (Mithen 1996, p. 132). The largest difference between the Early Human mind found in *Homo erectus* and the one found later in the Neanderthals, according to Mithen, is due to the development of linguistic intelligence, as evidenced by their possession of a modern hyoid bone in the vocal tract that is crucial for speech production, and by the modern size of their frontal cortex (Mithen 1996, p. 142).

Mithen admits to an apparent contradiction in regard to the social intelligence that he attributes to Early Humans. Neanderthals appeared to have lived in smaller groups than would be predicted by their brain size. Archaeological analysis of their living sites also reveals little evidence of social structure (Mithen 1996, p. 136). Here Mithen contravenes the archaeological evidence on the basis of psychology. He claims that Neanderthal social intelligence did not appear in the fossil record because the execution of technical tasks such as the butchering of carcasses was not integrated with their social life, so that the sites containing the tools and the fauna were not actual living sites. Mithen claims that the late version of the Early Human mind, as found in the Neanderthals, is fully modern in each of its component intelligences, but that it is different from the fully modern human mind in that it shows no evidence of cognitive fluidity between these domains.[32]

Anatomically modern humans, *Homo sapiens sapiens*, are frequently held to have emerged from *archaic Homo sapiens* in southern Africa (e.g., Klassies River Mouth and Border Cave) over 100,000 years ago and to have appeared in the Near East at Skuhl and Qafieh at 80,000–100,000 years ago, during the Middle to Upper Palaeolithic transition. The "cultural explosion" did not follow (according to Mithen) until 40,000 years later. Bone tools became widespread at this juncture and Mithen argues that this was possible because the modern mind could integrate technical and natural historical forms of intelligence. In fact, what happens with the emergence of the modern mind is the full integration of the various forms of previously modularized cognition (Mithen 1996, p. 153). Mithen identifies this form of cognitive integration with the type of 'meta-representational module'

---

[32] It is interesting to note that Fodor claims that it takes a synthetic integrated general intelligence to see that things you know about cows (natural intelligence) and things you know about fires (technical intelligence) can be used to conceive of how to make steak au poivre (Fodor 1998, p. 159). I believe we can therefore assume that the Neanderthal menu was quite impoverished compared to that of Cro-Magnon.

hypothesized by Sperber (1994) and with the seamless integration of multiple intelligences proposed by Gardner (1983). The type of cognition that makes artistic representation and religious belief possible is the ability to map across different domains (Carey and Spelke 1994) or, in other words, to transform coordinates from one conceptual space into another (Boden 1990).

Mithen also discusses artistic representation in general. He argues that the carved figurines and cave paintings that began to appear in Europe about 35,000 years ago (and only a little later in southern Africa) demonstrate that the prehistoric mind was fully modern in its synthetic capacities by this time. Worldwide differences in the onset time of these artistic activities may be due to variation in ecological and economic variables from one locale to another (Boden 1990, pp. 156–157). The depicted and sculpted figures typically involve humans that are part animal or humans that have animal traits, and the ability to fuse these unlike elements into single representations suggests that anthropomorphism and totemic forms of thought necessarily map between natural historical and social domains, while the sheer ability to produce such representations includes the technical domain.

It is important to note that Mithen follows Pfeiffer (1982) in interpreting the purpose of prehistoric art. Despite a minor level of acknowledgment that group identity and status were artistically represented, the social aspect of art seems to have been largely subordinate to the natural historical aspect. Mithen and Pfeiffer believe that artwork served as a storage bank for natural historical data. Such repositories of knowledge about animals helped prehistoric hunters hunt. Although anthropomorphism is incorrect in terms of animal cognitive psychology, attributing human folk psychology to game helped hunters predict animal behavior and thus bring home the kill.

What then can be said about language in Mithen's theory of cognitive evolution. Its role in the mind of Early Humans has only been accorded small aspect within the 'chapel' of social intelligence. However, in anatomically modern humans it becomes the portal through which nonsocial information invades social intelligence and becomes the foundation on which Mithen's 'cathedral'[33] of fluid consciousness is built. In other words, language forms the super-chapel of the mind (Mithen 1996, p. 188).[34] Mithen claims that people gradually began to use language for reasons other than social communication. Once technical and natural historical data could be represented via language, reflexive self-consciousness emerged and the 'cultural explosion' occurred. These developments coincided with the working of bone tool by anatomically modem humans approximately 50,000 years ago.

---

[33] Mithen uses the metaphor of a cathedral to characterize the existence of an un-encapsulated mental region. Please note that neo-Platonists always seem to end up in church at the end of the day.

[34] Mithen's view of language and conciousness is modeled after Sperber's concept of the 'meta-representational module' (Sperber 1994).

In the last chapter of his book, Mithen draws a helix of the evolution of consciousness (1996, p. 211). The common ancestor is equipped with general intelligence and an already highly developed social intelligence bequeathed by 35 million years of primate social manipulation. The course of evolution for nearly 2 million years from early Homo to Early Humans was facilitated through the development of specialized cognitive functions devoted to tool manufacture and subsistence resource management. This was followed by a phase in evolution characterized by a new kind of general intelligence, made possible by language, where the domain-specific Pleistocene 'Swiss Army-knife' intelligence of Cosmides and Tooby (1994) was replaced by cognitive fluidity (Karmiloff-Smith 1992). Mithen hypothesizes that the oscillation back and forth between generalization and specialization was necessary for the integrative evolutionary spiral that ended in cognitive fluidity. The limits of general intelligence led to the need for specialized intelligences. When these went as far as they could go on their own, then the only way forward was via integration. Hence, Mithen offers us a picture of the evolution of the mind where we evolve from a Tooby and Cosmides modular encapsulated 'jack-knife' into a Fodor 'language of thought' general processor. Important evolutionary theorists such as Maynard-Smith and Szathmary (1999, p. 143, 165) agree that this cognitive revolution was one of the fundamental shifts in the history of human evolution.

Weak Versus Massive Modularity

At this point, we have painted a reasonably detailed picture of Mithen's theory of prehistoric cognitive architecture. Are there any cracks in the walls of his theoretical 'cathedral'? I believe the biggest crack can be found in the concept of general intelligence. Those who follow Thurstone (1938), such as Gardner (1983), do not think there is any such thing as general intelligence. The strongest criticism can be found in Posner (1985) in reaction to Jensen's (1985) g-factor proposal. Briefly, Posner shows that the purported g-factor is so highly correlated with reading speed that one has to conclude that we are really looking at an aspect of verbal intelligence which can be referred to as verbal fluidity. In fact, proponents of the multiple intelligence view hold that there are only specialized intelligences and that general intelligence is a chimera. The research summarized by Gardner shows that people's abilities on these multiple dimensions of intelligence vary independently of one another. The question is whether there is something else called g-factor in addition to them. The answer appears to be negative.

One area of research that supports the view that intelligence is multiple rather than singular comes from behavioral genetics. Employing the classic twin study paradigm, Hughes and Plomin (2000) found that genetically identical twins showed a statistically significant higher correlation between both verbal intelligence and 'Theory of Mind' (which is a form of social intelligence) than did fraternal twins (2000, p. 55). The found that genetic factors accounted for approximately 2/3 of the variance in 'Theory of Mind' performance; but they also determined that this variance was independent of that for verbal intelligence (2000, pp. 56–57). One-third of the variance in 'Theory of Mind' performance was accounted for by non-shared environmental factors, with shared environment playing a negligible role. On the other hand, about half the correlation between verbal intelligence and 'Theory of Mind' is attributable to genetic factors while the other half of the correlation is attributable to shared (rather than non-shared) environment (2000, p. 58). Hughes and Plomin argue that because shared and non-shared environment affect verbal and social intelligence differentially the two are distinct modules (2000, 61). Because the environmental influences act in contrary ways with respect to the two modules, Hughes and Plomin also conclude that, despite the strong genetic component for each, there is strong input from the environment and that therefore Karmiloff-Smith's view of emergent modularity is supported against the hard-wired view offered by Fodor (2000, p. 60).

If the multiple intelligence view proves to be correct, what consequences follow for Mithen's theory? First, it would be inappropriate to characterize chimpanzees or our shared common ancestor as possessing a low level of general intelligence rather than just low levels of both technical and natural history intelligence. Second, the mind of modern humans could still be, as Cosmides and Tooby would have it, a 'Swiss Army-knife.' The latter is not just something that came and went with the Pleistocene. The most impressive piece of research in support of the domain-specific view has been Cosmides and Tooby's research on social cheating in reference to the Wason task. Their experiment put evolutionary psychology on its cognitive feet. Fodor (2000) attempts to strike this target specifically to undermine evolutionary psychology's *raison d'etre*. Fodor argues that the per-formance differential on the Wason task could be a function of the difference between the inferential paths that distinguish indicative from deontic conditionals, i.e. statements that assert truths versus statements that impose obligations (2000, p. 102). That is, the result is explainable as a feature of logic rather than evolu-tionary psychology. Fodor demonstrates his argument with extensive use of Ps and Qs. Either

(1) if P is under 18, Q s/he is drinking coke.
or
(2) If P is under 18, Q it is *required* that s/he drink coke.

Only subjects who are aware that the assertion involves an obligation always remember to check the age of the Ps who are not drinking coke. While Cosmides and Tooby argue that we use different parts of our mind to evaluate these two types of sentences and we are better at deontic logic, Fodor argues that the Wason task is set up in a logically confusing way and the task is only clear when the obligation is stated explicitly. Sperber et al. (1995) and Sperber and Girotto (2003) also criticize Cosmides' and Tooby's paradigm. They propose that *relevance theory* explains the poor performance of subjects on the indicative version of the Wason task. They argue that subjects do poorly on the indicative task because of the arbitrary nature of the logical connection between the p and the q (i.e. whether or not a card showing a vowel has an even number on the reverse side. They hypothesized that additional background information would improve the performance on the indicative so that it would match the deontic condition. Their experiments did indeed demonstrate that one could undermine the test by altering the wording of the questions so that the cues were either more or less intelligible to the respondent. They even showed that it was possible to reverse the outcome by changing the language used to present the task. Respondents would fail to correctly identify the correct response in the deontic condition and correctly identify the cards that need to be turned over in the non-moral condition.

If it is true that Cosmides' and Tooby's results are artifactual and thus do not provide evidence for a modular form of encapsulation, how damaging would it be to the domain specific view of cognition? It could prove devastating. Buller (2005), who provides a comprehensive review of research on the cheater-detection module, concludes that there is no cheater-detection module. Moreover, Buller argues on the basis of these problems with the Wason task, that the mind is domain-general rather than domain specific (Buller 2005, pp. 195–200). Sperber and Girotta, on the other hand, hold that it is still possible to defend the existence of a cheater-detection adaptation. They break the cheater-detection module into two independent conceptual primitive components that consist of rights and duties which can be used to evaluate the actions of others in a way that does not entail the pre-existence of a social contract concept and a disposition to check contracts for falsifying instances (Sperber and Girotta, p. 221). In other words, Sperber and Girotta believe that the mechanism for cheater-detection does not require the existence of an evolved module that is strictly devoted to cheater-detection and cheater-detection alone.[35] Nevertheless, the strongest and simplest evidence they cite in this regard comes from Mealy et al. (1996) who found that individuals spend significantly more time examining the faces of purported cheaters than non-cheaters. The latter finding supports the hypothesis that individuals are emotionally disposed to protect themselves from being exploited by others. Hence, while Sperber and Girotto are willing to countenance a cheater-detection adaptation, which could be described as a mechanism, they think it would be erroneous to grant such a mechanism the status of a stand-alone module. Possibly, a cheater-detection mechanism could be conceived as a subset of a more general domain such as social intelligence.

---

[35] Unfortunately, Sperber and Girotta formulate their critique of Cosmides and Tooby in the idiom of propositional-attitude psychology. This results in a lot of "I believe that she believes that he believes that I believe" formulations that are unwarranted. This sentence happy model of mental representation was originally advanced by Sperber and Wilson (1986) and further refined in the 2nd edition (Sperber and Wilson 1995).

Modularity and the Theory of Multiple Intelligence

I believe that multiple intelligence theory might also be interpreted to offer support for a more modest form of modularity for cognitive functions.[36] With respect to representational cognition, therefore, I suggest that we focus on studies that show how brain damage affects some domains of cognition while leaving others unaffected. A good example of this can be found in Steklis and Kling (1985), in which they demonstrate how social affiliation in nonhuman primates is affected by damage to specific and localized neural circuitry. They referred to this circuitry as a 'social affiliation' module prior to the official debut of evolutionary psychology. Damasio discusses clinical cases where brain damage to the ventromedial region of the prefrontal cortices of several human specimens selectively impaired social and intrapersonal intelligence, leaving other functions intact (1994 pp. 34–43). Damasio provides an extensive review of the clinical literature on case studies involving brain damage and on animal studies involving the ablation of neural tissue and proposes a neurobiological model for understanding the functional differentiation of the brain-mind. However, since evolutionary psychology has opted for a cognitive perspective, very little attention has been paid to the potential relevance of neurobiological perspectives.

The neurobiological approach to module specification provides the basis for a weaker and more developmentally labile perspective on the localization and structure of cognitive functions. I propose that we draw a distinction between weak and strong versions of the modularity thesis. According to Fodor, central cognitive processes are not modular because they are not fully encapsulated the way that input processes are. That is, they must be totally isolated from one another and impervious to training to be called modular. This is an unnecessarily strong thesis. The weaker version of the thesis only needs to argue that higher brain function consists of multiple independent processing systems as hypothesized by the hemispheric specialization theory or the multiple intelligence theory. The answer to any questions about the modular nature of such thinking depends on whether you accept Fodor's strong modularity thesis or some weaker thesis that permits different parts of the brain to work together separately on different aspects of a task. In this case, the massive modularity thesis of the evolutionary psychologists is intelligible as long as it is aligned with the weak version of the modularity thesis. That is to say, all cognitive systems are modular, not just perceptual input systems, but cognitive systems are only weakly modular. Input systems may demonstrate a stronger form of modular organization than cognitive systems, but they too should also be expected demonstrate some degree of developmental contingency.

What then of Mithen's use of the notion of cognitive fluidity? Cognitive "fluid" may be like the cosmic "ether" that physicists used to believe in but abandoned

---

[36] I argued in Chap. 2 that emotional circuitry is modular and Buller would seem to agree (2005, p. 151). Recall that Fodor, on the other hand, does not include emotional processing in the pantheon of cognition and so would find arguments about emotional modularity irrelevant.

when it was demonstrated that it was an unnecessary additional hypothesis. The multiple intelligence view maintains that some tasks require more than one type of intelligence. Catching a fly-ball in a baseball game, for example, would require spatial as well as kinesthetic intelligence. Singing a song would require musical as well as verbal intelligence. According to Gazzaniga (1985), hemispheric special-ization research shows that different aspects of cognitive performance are inde-pendently controlled by specialized modules located in different parts of the brain. These 'modules' are simultaneously active as dictated by the components of the tasks that are being performed. Gazzaniga also argues that this type of modular-ization is also present in normal humans. This does not mean that the different 'modules' do not act in concert. In fact, they are each performing their own special functions like the different instruments in an orchestra. According to Damasio, the intuition that mental activity is integrated and occurs in one place is an illusion produced by the 'parcellated' activity of many separate neural systems. 'Parcel-lated' refers to the fact that separate systems are active simultaneously (Damasio 1994, pp. 94–96).

The language module consists of different subprocessors that perform different functions, which relate to different aspects of linguistic representation. Gazzaniga maintains that a human is capable of building an independent linguistic repre-sentation of behavior that is generated by nonverbal 'modules.' This seems to correspond to Sperber's (1994) 'module of meta-representation.' Sperber and Gazzaniga's concept of 'modularity' is thus distinct from the kind of 'modularity' proposed by Fodor (1983). But the concepts do not compete for the same territory. Fodor's modular processes apply only to sensory input processes. On this view, Fodor's unencapsulated 'language of thought' is the type of thinking that goes on in Gazzaniga's language module and in Sperber's meta-representational module. Social, technical, and natural historical relationships are meta-representations created via language that occurs in various combinations. Hence, instead of con-ceptualizing language as a portal through which nonsocial intelligence flows into social intelligence, all the cognitive domains can continue to stand their ground as separate domain-specific processing centers.[37]

As far as the argument from developmental plasticity goes, critics of the modularity thesis such as Buller (2005) ascribe one-sided power to the role of the environment in the construction of the nervous system. Marcus (2004) has clearly explicated the roles that genes and environment play in the establishment of neural systems, and genes play a highly specific—rather than indeterminate—role in the construction of such systems. In some ways, developmental plasticity appears to be a failsafe mechanism to counteract local variation in the developmental context. It helps to preserve adaptation rather than undermining it. Hence developmental plasticity is no argument against modularity. As such, it is no surprise that research

---

[37] This not to imply, however, that processing centers involved in human language are conducted in a kind of language-of-thought as proposed by Fodor (1975; 2008). I maintain, as I argued in Chap. 1, that some variant of connectionist AI is capable of modeling the neurobiology of language (see Churchland and Churchland 1998).

on hemispheric lateralization demonstrates that there is some modification of the neural systems that underlie various areas of cognitive function during ontogeny (e.g., de Schonen and Mathiver 1989). Nevertheless, the end product is a domain-specific cognitive architecture. Thus, Mithen's metaphor of the mind as a cathedral that contains multiple chapels that flow into each other in the modern mind is misleading. Distinct modules need not flow into each other.

Language probably had the greatest influence on the prehistoric emergence of cultural behavior, since analogical reasoning seems to be the Rubicon. As Lakoff and Johnson (1980) have shown, language is inherently metaphorical. It is almost impossible to say anything without employing metaphors and analogies that make abstract relationships concrete and visual. I suspect that the necessity of employing metaphors in speech led to pictorial and sculptural means of representation as well, although it is also possible that the latter came first, followed by linguistic symbolism. The representation of relationships in one domain with the help of terms drawn from a different domain is almost certainly the basis of metaphor, including art and other forms of culture such as religious belief. In fact, metaphor and analogy are necessary because different cognitive domains are separate and do not communicate freely (cf. Gentner 1983). Social things or relationships can be represented in terms of natural things or relationships, and the differences can help make the similarities clearer, or the similarities can help make the differences clearer. Specific attention to these aspects of neuropsychology and neurolinguistics would have eliminated the need for Mithen to mix different competencies together via the concept of cognitive fluidity.

Explaining the Evolution of the Pleistocene Mind–Brain

Much of Jerry Fodor's invective against evolution is directed at the "just so" stories offered by sociobiologists and evolutionary psychologists. Fodor attacks Dawkins (cf. Fodor 1998, pp. 163–170) and Pinker (cf. Fodor 1998, pp. 203–214) as "just so" theoreticians, and it is clear that he has thrown his lot[38] in with hostile critics of adaptationist thinking such as Eldredge (1995) and Gould and Lewontin (1979). Yet behavioral ecology and paleoanthropology could help these adaptive arguments become more empirical. For example, although Fodor doubts that the 'selfish gene' theory can explain why parents love their children, preferring the non-Darwinian explanation that parents just love their children because they love them (Fodor 1998, pp. 211–212)—the 'expensive tissue' hypothesis (Aiello and Wheeler 1995; Key and Aiello 1999) goes far in explaining the evolution of parental care in hominids.

---

[38] Pun intended.

The 'expensive tissue' hypothesis holds that cooperative breeding, grand-mothering, extensive cooperation with nonkin (i.e., reciprocal altruism), food sharing, theory of mind, and guilt are necessary elements of human social organization. Their fundamental thesis about the evolution of human social cooperation is stated thus: "As the energetic cost of reproduction increases, so does the like-lihood of cooperation between females and also cooperation between females and males." (Key and Aiello 1999, p. 18) Reproductive costs to females include the production of gametes, gestation, lactation, and childcare, while mating effort and mate guarding comprise the reproductive costs to males. Cooperation between females in many mammalian species is widespread in the form of alloparenting (e.g., babysitting, suckling). Reciprocal altruism is mutually beneficial between females because they tend to have higher energetic costs. The fossil record shows that in australopithecines, sexual body dimorphism is such that males were 40% larger than females, suggesting that males had to compete against each other for mates (mating effort). In early *Homo*, on the other hand, body dimorphism is lower. Males were only 20% larger than females. It has been argued that this demonstrates social cooperation between males instead of competition (Foley 1987).

Increase in brain size appears to be crucial when approaching the Rubicon of cultural influence. Key and Aiello discuss the factors contributing to encephali-zation in *Homo*. *Homo* shows a threefold increase in brain size compared to the australopithecines. At the same time, Homo showed a decrease in the energetic investment in the gastrointestinal tract. It became smaller, it is argued, because meat is easier to digest than the hard tubers that australopithecines ate. Further-more, the increase in brain size for *Homo* implies a higher energy cost to females in terms of gestation and increased levels of infant care due to longer infant dependency. Key and Aiello note that this increase resulted in selection pressure for alloparenting by females towards other females. Thus, 'grandmothering' was made possible by the evolution of menopause and longer life spans after menopause.

Key and Aiello argue that the increase in encephalization during the Paleolithic between 500,000 and 100,000 years ago coincides with the appearance of large kill sites and the increased reliance on meat in the hominid diet. Hunting by males was an important source of protein for hominid females. During this period, this was probably the major form of male parental investment. Following Hawkes (1993), they compare provisioning in early *Homo* to that of contemporary hunter-gatherers. In hunter-gatherer societies such as the Ache and Hadza, as much as 85% of calories are provided via hunted animal protein. They hypothesize a dual pattern based on age and marital status. First, young unmarried men engage in high risk hunting which, when successful, provides the group with a windfall of valuable meat. Their success at hunting is converted into reproductive success; they gain status and mates because of it. Second, older married men engage in lower risk hunting, acting as providers to their families and neighbors. It is important to note that young men do not dispose of their kill individually. It is alienated to the group, or, after they are married, to the exogamous group that they married into as a form of "brideservice." The meat becomes part of the system of alliance formation and maintenance.

Hence, departures from the australopithecine social prototype occurred because the hominids that comprised early *Homo* moved into the savanna where the acquisition of animal protein became increasingly important. Because of the patchy distribution of animal protein, food sharing and social cooperation between males was rewarded with reproductive success, while the cost of reproduction to females rose as a result of brain expansion and infant dependency. The latter costs could also be buffered by maintaining ties with kin outside one's natal community and by establishing ties of reciprocal altruism with non-kin for purposes of acquiring alloparenting. Affinal ties made possible by the development of enduring mating relationships provided males with valuable alliances for hunting and for warfare against outside groups. Key and Aiello maintain that humans became better at remembering complex social relationships, and that they also evolved the ability to exploit these relationships over time and space in ways that contributed to inclusive fitness. Conversely, individuals also become better at protecting themselves from cheaters (i.e. non-reciprocators). The undeniable conclusion that follows from the preceding analysis is that the evolution of the mind–brain is explained by specific selection pressures during the Pleistocene that bear on social and familial cooperation that clearly distinguish it from that of Australopithecine evolution.

In the final analysis, Fodor's anti-adaptationist stance is genuinely puzzling since the form of psychological explanation that he purveys exhibits the same adaptationist commitment, albeit applied to different type of phenomenon. As Kalke (1969) and Causey (1977, pp. 142–151) pointed out long ago, Fodor's anti-adaptationist view is predicated on a commitment to functionalism, which stipulates that physical states are mere tokens of functional types that exist at a higher level. As I argued in Chap. 1, the word "level" implies a distinct onto-logical "level" for functional entities; this "level" is where Fodor thinks psychological explanation belongs. For example, to Fodor and other cognitive functionalists the physical differences between cats and mechanical mousetraps are insignificant since they both perform the function of catching mice. It is this same form of explanation that got sociobiologists in trouble with Gould and Lewontin (1979) or Vayda (1995). These critics question the thesis that the various means one uses to achieve the functional goal of maximizing one's reproductive fitness are insignificant, and that it is the end goal that determines the need for a strictly functional rather than a proximate or physical explanation. Perhaps it is time for Jerry and his granny to turn their spears around and also time for evolutionary psychologists to seek assistance from empirical disciplines such as cognitive neurobiology rather than pursuing the latest version of the Cartesian nervous system offered by philosophers and evolutionary psychologists who remain wedded to propositional-attitudes and the language-of-thought paradigm. Those committed to the latter always come down on the side of anti-materialism, anti-reductionism, and antievolutionism.[39] Although, Cartesian dualism may offer comfort in form of the illusion of the immateriality of the 'soul' or the existence of free will, these illusions are inimical to scientific progress. The latter requires an unwavering commitment to philosophical materialism.

---

[39] It is little wonder that Fodor recently attempted to sink the entire Darwinian flagship. See Fodor and Piattelli-Palmarini (2010). Their destroyer, however, was torpedoed immediately upon leaving port by Block and Kitcher (2010) who deftly explained the fundamental misunderstandings of Darwinism that were leveled in the Fodor/Piattelli-Palmarini critique. I submit that Fodor's underlying motivation in the critique is to preserve Platonism from being parted out through Darwinian evolution, since the differential reproduction of genetic materials over multiple generations would necessarily decimate the Platonic wholes that Fodor needs to preserve in order to sustain his own highly Platonic position. (Dennett incisively refers to the implications of Darwinism for Platonism as 'Darwin's Dangerous Idea' (Dennett 1995). The full-bodied Platonist will eventually turn on any position that is less than the fully committed version, and so a Piagetian-Lamarckian such as Piattelli-Palmerini will be cast off in a leaky boat as soon as the common enemy, in this case Darwin, is thought to be successfully torpedoed.

Epilogue

What Could an Evolutionary Theory of Mind be About?

Eliminative materialists such as Paul and Patricia Churchland argue that the 'theory of mind' offered by folk psychology is radically wrong. They argue this because the theoretical terms of folk psychology do not, and cannot, model the kinematics and dynamics of real neurobiological systems. If I have succeeded, I have convinced you that they are correct in this diagnosis. The question then becomes, "Whither folk psychology?" The Churchlands argue that two possible options exist:

(a) We can attempt to *reduce* the terms of folk psychology to those of neurocognitive science.
   Or
(b) We can *replace* folk psychology with a paradigm that eliminates the terms of folk psychology from psychological explanation.

On behalf of the first option, Churchland and Churchland (1998, pp. 65–80) present the case for reduction by comparing the situation to other cases in the history of science where such reduction was obtained, for example, the reduction of Kepler's Laws to Newton's, and, in turn, the reduction of Newton's Laws to those of Einstein's Special Theory of Relativity. Another example is the reduction of classical chemistry to that of quantum physics. The Churchlands make the case that it might be, at least theoretically, possible to effect a smooth reduction of folk psychology to neurobiology.

The primary problem that obtains with the reductionist alternative devolves on the issue of the incommensurability of competing theories. Is it possible to call it a case of reduction when the theoretical terms of one theory apply to completely different phenomena than those of the second theory? This issue was raised by Thomas Kuhn in *The Structure of Scientific Revolutions* (1962). The ensuing debate in philosophy of science only exacerbated these concerns (cf. Lakatos and

A. Walter, *Evolutionary Psychology and the Propositional-attitudes*,
SpringerBriefs in Philosophy, DOI: 10.1007/978-94-007-2969-8,
© The Author(s) 2012

Musgrave 1970; Suppe 1977).[40] If we're not talking about the same thing, how can we speak of reduction? This appears to be the case when we try to reduce mental events to physical brain events or, alternatively, to construe both as twin aspects of one phenomenon. The attempt to map single 'mental' events onto single brain events via the identity theory results in a species of dualism (Feyerabend 1963b). The attempt to match one 'mental' event to more than one physical realization (e.g. Kim 1986) is also dualistic.[41] The problem, as outlined by Donald Davidson (1963)—and driven home by Quine (1985)—is that the terms of folk psychology and scientific psychology are anomalous. This is so because they are not working the same side of the street. (In fact, they are not even working the same street). Even though we know (unless we accept dualism) that there are only the physical events of the brain, we use mentalistic language in casual discourse. The question then becomes, should science remain prisoner to folk psychological conceptions or go its own separate way? Behaviorists, of course, chose the latter path by eliminating internal variables. Behaviorists, as it were, thus form a species of eliminativist materialist. Still, what about the 'inside' story? Can cognitive neurobiology replace folk psychology?

Churchland (1981) has also presented the case for replacement, arguing that the terms of folk psychology are so at odds with neurobiological explanation that the former must be abandoned altogether. The view that I have defended in these essays is that this indeed must be the case for the affective neurosciences and for cognitive neurobiology. I believe that a mechanist theory of motivational endowments and cognitive representation must replace folk psychology in the scientific explanation of behavior. Some, such as Stich (1996, pp. 34–35), who follows Lycan 1988), argue that all we have to do is switch the referent in order to retain the term. As Lycan showed Stich, we now know that stars are not holes in the sky through which heavenly light shows, but instead are massive luminous balls of plasma held together by gravity. We have switched the referent of the term. Stich believes we can do the same for terms inherited from pre-scientific folk psychology. That is, the terms of folk psychology could be preserved; but instead of referring to beliefs or desires, as such, they could be recast to refer to adaptations or other subdoxastic psychological mechanisms. Now, although this may prove to be a useful strategy for some pre-scientific terminology, I would not want to prejudice the future of psychological explanation by stipulating that the best way to advance is by finding mechanistic referents for folk terms. Indeed, I think the language of adaptive mechanism will necessarily displace the language of ordinary folk psychology in scientific discourse. And I think evolutionary psychology is well placed because of its concern with adaptive mechanisms to help

---

[40] At which point the issue strangely receded into the woodwork without resolution. The sociology of knowledge appears to have cannibalized philosophy of science's concern with the topic (Fuller 1992).

[41] Some thinkers hold these two views simultaneously. Compare the Chap. 1 of John Searle's Minds, Brains, and Science (1984) to the Chap. 4.

redirect the language in which representational and motivational mechanisms are described.

The very existence of the goal of achieving a scientific explanation of adaptation, however, raises another explanatory issue for scientific psychology, but I think it is one that evolutionary psychology is well equipped to face. This issue is the question of why, if folk psychology is so radically wrong as the Churchlands claim it is, and I agree that it is radically wrong, how do we explain the existence of folk psychologies? The answer also bears on the question as to whether we can eliminate folk psychological concepts from our species' psychology as the Churchlands' hope. Now the answer to the question of the existence of folk psychology could possibly be chalked up to human ignorance, as could be the case in pre-scientific folk-astronomy or folk-zoology; we might therefore expect folk psychology to disappear with improvements in our scientific understanding of behavior and the mechanisms that produce it. However, evolutionary psychologists have already undermined this ambitious hope. The answer to the question of why folk psychology exists in the first place also explains why we will continue to think of ourselves in its terms—despite the metaphysically incorrect nature of that understanding.

Evolutionary psychologists who work in the area of 'theory of mind' (i.e. ToMM) take as their research domain the discovery of why we attribute certain beliefs and desires to others and to ourselves. This was the focus of Byrne and Whitten (1988) and others who investigated 'Machiavellian' intelligence, and it is also the research domain of others such as Gelman (2003) who've studied the development of essentialist ideas in children. ToMM (Leslie 1994, 2011) has as its proper domain the investigation of the evolution of folk psychological concepts. Essentialism may be false as Wittgenstein (1953) and others have demonstrated, but an account of why we think in such terms is an account worth pursuing. The account has both evolutionary historical and proximate developmental aspects. We evolved to think in such terms, not because they give us an accurate picture of our own inner machinations, but because they worked in terms of helping us to predict and control the behavior of others in ways that facilitated differential reproductive success in ancestral environments.[42] What evolutionary psychologists provide, therefore, is an evolutionary explanation of the practical virtues of successful communication. Therefore, even if we cannot eliminate folk psychology from psychological explanation, as envisioned by eliminative materialists such as the Churchlands, we can at least move the terms of folk psychology from *explanans* to *explanandum*. This, and not a theory of mental representation, is the proper goal of the evolutionary psychologist.

---

[42] See Aiello and Dunbar (1993) or Tomasello (2008) for accounts of the evolutionary context in which these communicative competencies were selected.

References

Aiello, L. C., & Dunbar, R. (1993). Neocortex size, group size, and the evolution of language. *Current Anthropology, 34*, 84–93.

Aiello, L. C., & Wheeler, P. (1995). The expensive tissue hypothesis: the brain and the digestive system in human evolution. *Current Anthropology,36*, 199–221.

Aldridge, K. (2010). Patterns of differences in brain morphology in humans as compared to extant apes. *Journal of Human Evolution,60*(1), 94–105.

Allen, M. (1983). Models of hemispheric specialization. *Psychological Bulletin,93*, 73–104.

Ansari, D. (2011). Introduction to the special issue: Toward a developmental cognitive neuroscience of numerical and mathematical cognition. *Developmental Neuropsychology,36*(6), 645–650.

Aquinas, T. St. (1955). *Summa Theologica* (pp. 1265–1274). Chicago: Encyclopedia Britannica.

Armstrong, E. E. (1992). The limbic system and culture: An allometric analysis of the neocortex and limbic nuclei. *Human Nature,2*(2), 117–136.

Arnold, M. (1984). *Memory and the brain*. Hillsdale, NJ: Lawrence Erlbaum & Associates.

Baker, R. R., & Bellis, M. (1995). *Human sperm competition*. London: Chapman and Hall.

Barash, D. (1977). *Sociobiology & behavior*. New York: Elsevier.

Belluzzi, J. D., & Stein, L. (1977). Enkephalin may mediate Euphoria and drive-reduction reward. *Nature,266*, 556–558.

Bickerton, D. (1990). *Language and species*. Chicago: University of Chicago Press.

Binford, L. R. (1985). Human ancestors: Changing views of their behavior. *Journal of Anthropological Archaeology,4*, 292–327.

Block, N. (2007). *Consciousness, function and representation*. (Collected papers, Vol. 1). Cambridge, MA: MIT Press.

Block, N., & Kitcher, P. (2010). Misunderstanding Darwin: Review of Jerry Fodor and Massimo Piatell-Palmarini. What Darwin got wrong. *Boston Review*, March/April, 1–9.

Blumenschine, R. J. (1987). Characteristics of an early hominid scavenging niche. *Current Anthropology,28*, 383–407.

Blurton Jones, N. G. (1990). Three sensible paradigms for research on evolution and human behavior. *Ethology and Sociobiology,11*, 353–359.

Boden, M. (1990). *The creative mind: Myths and mechanisms*. London: Weidenfeld & Nicolson.

Bolles, R. C., & Fanselow, M. S. W. (1982). Endorphins and behavior. *Annual Review of Psychology,33*, 86–101.

Buckner, R. L., Andrews-Hanner, J. R., & Schacter, D. L. (2008). The brain's default network: Anatomy, function, and relevance to disease. *Annals of the New York Academy of Sciences,1124*, 1–38.

A. Walter, *Evolutionary Psychology and the Propositional-attitudes*,
SpringerBriefs in Philosophy, DOI: 10.1007/978-94-007-2969-8,
© The Author(s) 2012

Buller, D. (2005). *Adapting minds: Evolutionary psychology and the persistent quest for human nature*. Cambridge, MA: MIT Press.

Buss, D. M. (1991). Evolutionary personality psychology. *Annual Review of Psychology,42*, 459–491.

Buss, D. M. (1994). *The evolution of desire: Strategies of human mating*. New York: Basic Books.

Buss, D. M. (2005). *The handbook of evolutionary psychology*. Hoboken, NJ: John Wiley & Sons.

Buss, D. M., & Reeve, H. K. (2003). Evolutionary psychology and developmental dynamics: Comment on Lickliter & Honeycutt. *Psychological Bulletin,129*, 848–853.

Byrne, R. W., & Whiten, A. (Eds.). (1988). *Machiavellian intelligence: Social expertise and the evolution of intellect in monkeys, apes, and humans*. Oxford: Clarendon Press.

Cantlon, J. F., Libertus, M., Pinel, P., Dehaene, S., Brannon, E., & Pelfrey, K. (2009). The neural development of an abstract concept of number. *Journal of Cognitive Neuroscience,21*(11), 2217–2229.

Carey, S., & Spelke, E. (1994). Domain-specific knowledge and conceptual change. In L. A. Hirschfield & S. A. Gelman (Eds.), *Mapping the mind: Domain specificity in cognition and culture* (pp. 169–200). Cambridge: Cambridge University Press.

Causey, R. (1977). *The unity of science*. Boston: D. Reidel.

Chalmers, D. (1996). *The conscious mind*. Oxford: Oxford University Press.

Chomsky, N. (1959). A review of B.F. Skinner's "verbal behavior". *Language, 35*, 26–58.

Chomsky, N. (1965). *Aspects of a theory of syntax*. Cambridge, MA: MIT Press.

Chomsky, N. (1972). *Language and mind*. New York: Harcourt, Brace & Jovanovich.

Chomsky, N. (1980). *Rules and representations*. New York: Columbia University Press.

Churchland, P. S. (1980a). Language, thought and information processing. *Nous,8*, 147–169.

Churchland, P. S. (1980b). A perspective on mind-brain research. *Journal of Philosophy,77*, 185–207.

Churchland, P. M. (1981). Eliminative materialism and the propositional attitudes. *Journal of Philosophy,78*, 67–90.

Churchland, P. M. (1982). Is thinker a natural kind? *Dialogue,21*(2), 223–238.

Churchland, P. S. (1983). Dennett's instrumentalism: A frog at the bottom of Dennett's cup. *Behavioral and Brain Sciences,6*, 358–359.

Churchland, P. M. (1986a). Some reductive strategies in cognitive neurobiology. *Mind,95*, 279–309.

Churchland, P. S. (1986b). *Neurophilosophy: Toward a unified understanding of the mind–brain*. Cambridge, MA: MIT Press.

Churchland, P. M. (1989). *A neurocomputational perspective*. Cambridge, MA: MIT Press.

Churchland, P. M. (2007). *Neurophilosophy at work*. Cambridge: Cambridge University Press.

Churchland, P. M., & Churchland, P. S. (1998). *On the Contrary: Critical essays 1987–1997*. Cambridge, MA: MIT Press.

Clark, A. (2008). *Supersizing the mind: Embodiment, action, & cognitive extension*. New York: Oxford University Press.

Cooper, L. A., & Shepard, R. N. (1973). Chronometric studies of the rotation of mental images. In W. G. Chase (Ed.), *Visual information processing* (pp. 75–176). New York: Academic Press.

Cosmides, L., & Tooby, J. (1992). Cognitive adaptations for social exchange. In J. Barkow, L. Cosmides, & J. Tooby (Eds.), *The adapted mind* (pp. 163–228). New York: Oxford University Press.

Cosmides, L., & Tooby, J. (1994). Origins of domain specificity: The evolution of functional organization. In L. A. Hirschfield & S. A. Gelman (Eds.), *Mapping the mind: Domain specificity in cognition and culture* (pp. 85–116). Cambridge: Cambridge University Press.

Cottrell, G. & Metcalfe, J. (1991). EMPATH: Face, emotion, and gender recognition using holons. In R. Lippman et al. (Eds.), Advances *in neural information processing systems* (Vol. 3, pp. 1–7). San Mateo, CA: Morgan Kauffman.

Culler, J. (1976). *Ferdinand de Saussure*. New York: Penguin.

Damasio, A. R. (1994). *Descartes' error*. New York: Putnam & Sons.

Davidson, D. (1963). Actions, reasons, and causes. *Journal of Philosophy,50*, 585–700.

Davidson, D. (1980). *Essays on actions and events*. Oxford: Clarendon Press.

Dawkins, R. (1976). *The selfish gene*. Oxford: Oxford University Press.

Dawkins, R. (1979). Twelve misunderstandings of kin selection. *Zeitschrift Fur Tierpsychology,51*, 184–200.

Dawkins, R. (1984). Replicators, consequences, and displacement activities. *Behavioral and Brain Sciences,7*(4), 486–487.

Dawkins, R. (2006). *The god delusion*. Boston: Houghton Mifflin Co.

de Schonen, S., & Mathiver, E. (1989). First come, first served: A scenario about the development of hemispheric specialization in face recognition during infancy. *European Bulletin of Cognitive Psychology,9*, 3–44.

Deacon, T. (1997). *The symbolic species: The co-evolution of language and the brain*. New York: W.W. Norton & Co.

DeMarest, W. J. (1983). Does familiarity necessarily lead to erotic indifference and incest avoidance because inbreeding lowers reproductive fitness? *Behavioral and Brain Sciences,6*, 106–107.

Dennett, D. (1983). Intentional systems in cognitive ethology: The 'panglossian paradigm' defended. *Behavioral and Brain Sciences,6*, 343–390.

Dennett, D. (1991). *Consciousness Explained*. Boston: Little, Brown.

Dennett, D. (1995). *Darwin's dangerous idea*. New York: Simon and Schuster.

Dennett, D. (1996). *Kinds of minds*. New York: Basic Books.

Dennett, D. (1998). *The logical geography of computational approaches: A view from the east pole. In Brainchildren: Essays on designing minds*. Cambridge, MA: MIT Press.

Dennett, D. (2003). *Freedom evolves*. New York: Penguin.

Derrida, J. (1973). *Speech and phenomena*. Evanston, IL: Northwestern University Press.

Dixson, A. F. (2009). *Sexual selection and the origin of human mating systems*. Oxford: Oxford University Press.

Driscoll, C., & Stich, S. (2008). Vayda blues: Explanation in Darwinian ecological anthropology. In B. Walters, et al. (Eds.), *Against the grain: The Vayda tradition in human ecology and ecological anthropology* (pp. 175–191). New York: Altamira.

Ekman, P. (Ed.). (1972). *Darwin and facial expression: A century of research in review*. New York: Academic Press.

Ekman, P., & Friesen, W. V. (1975). *Unmasking the face: A guide to recognizing emotions from facial expressions*. Englewood Cliffs, NJ: Prentice-Hall.

Eldredge, N. (1995). *Reinventing Darwin*. New York: John Wiley.

Elman, J. et al. (1996). *Rethinking innateness: A connectionist perspective on development*. Cambridge, MA: MIT Press.

Ellison, P., & Gray, P. (2009). *Endocrinology of social relationships*. Cambridge, MA: Harvard University.

Falk, D. (1983). Cerebral cortices of east African early hominids. *Science,221*, 1072–1074.

Feyerabend, P. (1963a). Materialism and the mind–body problem. *Review of Metaphysics,17*, 49–66.

Feyerabend, P. (1963b). Mental events and the brain. *Journal of Philosophy,60*, 295–296.

Flinn, M. V., Ward, C. V., & Noone, R. J. (2005). Hormones and the human family. In D. M. Buss (Ed.), *The handbook of evolutionary psychology*. Hoboken, NJ: John Wiley & Sons.

Fodor, J. (1975). *The language of thought*. New York: Crowell.

Fodor, J. (1983). *Modularity of mind*. Cambridge, MA: MIT Press.

Fodor, J. (1994). *The elm and the expert*. Cambridge, MA: MIT Press.

Fodor, J. (1998). *In critical condition: Polemical essays on cognitive science and the philosophy of mind*. Cambridge, MA: MIT Press.

Fodor, J. (2000). *The mind doesn't work that way: The scope and limits of computational psychology*. Cambridge, MA: MIT Press.

Fodor, J. (2008). *LOT 2: The language of thought revisited*. New York: Oxford University Press.

Fodor, J., & Piattelli-Palmarini, M. (2010). *What Darwin got wrong*. New York: Farrar, Strauss, Giroux.

Foley, R. (1987). *Another unique species*. New York: John Wiley & Sons.

Frank, R. (1988). *Passions within reason: The strategic role of the emotions*. New York: Norton.

Fuller, S. (1992). *Philosophy of science and its discontents*. New York: Guilford Press.

Gangestad, S. W., & Thornhill, R. (1997). The evolutionary psychology of extrapair sex: The role of fluctuating asymmetry. *Evolution and Human Behavior,18*, 69–88.

Gangestad, S. W., Thornhill, R., & Garver-Apgar, C. E. (2005). Adaptations to ovulation. In D. M. Buss (Ed.), *The handbook of evolutionary psychology*. Hoboken, NJ: John Wiley & Sons.

Gangestad, S. W., Thorhill, R., & Garver-Apgar, C. E. (2010). Fertility in the cycle predicts women's interest in sexual opportunism. *Evolution and Human Behavior,31*(6), 400–411.

Gardner, H. (1983). *Frames of mind: The theory of multiple intelligences*. New York: Basic Books.

Garver-Apgar, C. E., Gangestad, S. W., & Thornhill, R. (2008). Hormonal correlates of women's mid-cycle preference for the scent of symmetry. *Evolution and Human Behavior,29*(4), 223–232.

Gazzaniga, M. (1985). *The social brain: Discovering the networks of the mind*. New York: Basic Books.

Gelman, S. (2003). *The essential child: Origins of essentialism in everyday thought*. New York: Oxford University Press.

Gelman, S. A., Coley, J. D., & Gottfried, G. M. (1994). Essentialist beliefs in children: The acquisition of beliefs and theories. In L. A. Hirshfeld & S. A. Gelman (Eds.), *Mapping the mind*. Cambridge: Cambridge University Press.

Gentner, D. (1983). Structure-mapping: A theoretical framework for analogy. *Cognitive Science,7*, 155–170.

Globus, G. (1992). Toward a noncomputational cognitive neuroscience. *Journal of Cognitive Neuroscience.,4*(4), 299–310.

Gould, J. L., & Gould, C. G. (1988). *The honey bee*. New York: W.H. Freeman.

Gould, S. J. & Lewontin, R. C. (1979). The Spandrels of San Marco and the Panglossian paradigm. *Proceedings of the Royal Society of London Series B, 205*, 581–598.

Greenfield, P. M. (1991). Language, tools and brain: The ontogeny and phylogeny of hierarchically organized sequential behavior. *Behavioral and Brain Sciences,14*, 531–595.

Grice, P. (1957). Meaning. *Philosophical Review,64*, 377–388.

Griffiths, P. E. (1990). Modularity and the psychoevolutionary theory of emotion. *Biology and Philosophy,5*, 175–196.

Griffiths, P. E. (1997). *What emotions really are: The problem of psychological categories*. Chicago: University of Chicago Press.

Griffths, P. E., & Gray, R. D. (1994). Developmental systems and evolutionary explanation. *Journal of Philosophy,91*(6), 277–304.

Grossberg, S. (1988). Nonlinear neural networks: Principles, mechanisms, and architectures. *Neural Networks,1*(1), 17–61.

Hales, S. (2009). Moral relativism and evolutionary psychology. *Synthese,166*, 431–447.

Hauser, M., Chomsky, N., & Fitch, W. (2002). The faculty of language: What is it, who has it, and how did it evolve? *Science,298*, 1569–1579.

Hawkes, K. (1993). Why hunter-gatherers work: An ancient version of the problem of public goods. *Current Anthropology,34*, 341–361.

Heatherton, T. F. (2011). Neuroscience of self and self-regulation. *Annual Review of Psychology,62*, 363–390.

Hellige, J. B. (1990). Hemispheric asymmetry. *Annual Review of Psychology,41*, 55–80.

Hintzman's, D. (1990). Human learning and memory: Connections and dissociations. *Annual Review of Psychology,41*, 109–139.

Horgan, T., & Tye, M. (1986). Against the token identity theory. In E. LePore & B. McLaughlin (Eds.), *Actions and events: Perspectives on the philosophy of Donald Davidson* (pp. 427–443). London: Basil Blackwell.

Hughes, C., & Plomin, R. (2000) Individual differences in early understanding of mind: Genes, non-shared environment and modularity. In P. Carruthers & A. Chamberlain (Eds.), *Evolution and the human mind: Modularity, language and meta-cognition* (pp. 47–62). Cambridge: Cambridge University Press.

Humphrey, N. (1992). *A natural history of the mind*. London: Chatto & Windus.

Irons, W. (1990). Let's make our perspective broader rather than narrower. *Ethology and Sociobiology,11*, 361–374.

Issac, G. (1978). The food sharing behavior of proto-hominids. *Scientific American,238*, 90–108.

Izard, C. (1971). *The face of emotion*. New York: Appelton, Centruy, Crofts.

Jensen, A. (1985). The nature of black-white differences on various psychometric tests: Spearman's hypothesis. *Behavioral and Brain Sciences,8*, 193–258.

Joyce, R. (2006). *The evolution of morality*. Cambridge, MA: MIT Press.

Kalke, W. (1969). What's wrong with fodor and putnam's functionalism? *Nous,3*, 83–93.

Karmiloff-Smith, A. (1992). *Beyond modularity: A developmental perspective on cognitive science*. Cambridge, MA: MIT Press.

Karmiloff-Smith, A. (1994). Precis of beyond modularity: A developmental perspective on cognitive science. *Behavioral and Brain Sciences, 17*, 693–745.

Key, C. A., & Aiello, L. C. (1999). The evolution of social organization. In R. Dunbar, et al. (Eds.), *The evolution of culture* (pp. 15–33). New Brunswick, NJ: Rutgers University Press.

Kim, J. (1986). Psychophysical laws. In E. LePore & B. McLaughlin (Eds.), *Actions and Events: perspectives on the philosophy of Donald Davidson* (pp. 369–386). London: Basil Blackwell.

Kim, J. (2007). *Physicalism or something near enough*. Princeton: Princeton University Press.

Kitcher, P. (1985). *Vaulting Ambition*. Cambridge, MA: MIT Press.

Kitcher, P. (1989). Proximate and developmental analysis. *Behavioral and Brain Sciences, 12*(1), 186–187.

Kuhn, T. (1962). *The structure of scientific revolutions*. Chicago: University of Chicago Press.

Lakatos, I., & Musgrave, A. (Eds.). (1970). *Criticism and the growth of knowledge*. Cambridge: Cambridge University Press.

Lakoff, G., & Johnson, M. (1980). *Metaphors we live by*. Chicago: University of Chicago Press.

LeDoux, J. (1996). *The emotional brain*. New York: Simon & Schuster.

Leslie, A. M. (1994). ToMM, ToBY, and agency. Core architecture and domain specificity. In L. A. Hirchfeld & S. A. Gelman (Eds.), *Mapping the mind: Domain specificity in cognition and culture*. New York: Cambridge University Press.

Leslie, A. M. (2011). *The origins of mental representation*. New York: Wiley-Blackwell.

Lickliter, R. (2008). The growth of developmental thought: Implications for a new developmental psychology. *New Ideas in Psychology, 26*, 353–369.

Lickliter, R., & Honeycutt, H. (2003). Developmental dynamics: Towards a biologically plausible evolutionary psychology. *Psychological Bulletin, 129*, 819–835.

Lycan, W. (1988). *Judgment and Justification*. Cambridge: Cambridge University Press.

MacCorquodale, K. (1970). On Chomsky's review of Skinner's verbal behavior. *Journal of the Experimental Analysis of Behavior, 18*, 83–99.

Malatesta, C. Z., & lzard, C. E. (1984). The ontogenesis of human social signals: From biological imperative to symbol utilization. In N. A. Fox & R. J. Davidson (Eds.), *The psychobiology of affective development* (pp. 161–206). Hillsdale, New Jersey: Lawrence Erlbaum Associates.

Marcus, G. (2004). *The birth of the mind*. New York: Basic Books.

Marcus, G. (2008). *Kluge: the haphazard construction of the human mind*. New York: Faber & Faber.

Maynard Smith, J., & Szathmary, E. (1999). *The origins of life*. New York: Oxford University Press.

Maynard-Smith, J. (1982). *Evolution and the theory of games*. New York: Cambridge University Press.

McCarthy, J. (1979). Ascribing mental qualities to machines. In M. Ringle (Ed.), *Philosophical perspectives in artificial intelligence*. Atlantic Highlands, NJ: Humanities Press.

McLaughlin, B. (1986). Anomalous monism and the irreducibility of mind. In E. LePore & B. McLaughlin (Eds.), *Actions and events: Perspectives on the philosophy Donald Davidson* (pp. 331–368). London: Basil Blackwell.

McLaughlin, B., & Cotter, J. (Eds.). (2007). *Contemporary debates in philosophy of mind*. New York: Wiley-Blackwell.

Miller, G. F., & Todd, P. M. (1998). Mate choice turns cognitive. *Trends in Cognitive Science, 2*(5), 190–198.

Millikan, R. (1984). *Language, thought and other biological categories*. Cambridge, MA: MIT Press.

Millikan, R. (1993). *White queen psychology and other essays for Alice*. Cambridge, MA: MIT Press.

Minski, M. (2007). *The emotion machine*. New York: Simon & Schuster.

Mithen, S. (1996). *The prehistory of the mind: The cognitive origins of art & science*. New York: Thames & Hudson.

Morillo, C. (1990). The reward event and motivation. *Journal of Philosophy, 87*, 169–186.

Morillo, C. (1995). *Contingent creatures: A reward event theory of motivation and value*. Lanham, Maryland: Rowman & Littlefield.

Olds, J. (1977). *Drives and reinforcements: Behavioral studies of hypothalamic functions*. New York: Raven Press.

Oyama, S. (1985). *The ontogeny of information*. Cambridge: Cambridge University Press.

Panksepp, J. (1986). The neurochemistry of behavior. *Annual Review of Psychology, 37*, 77–107.

Panksepp, J. (1998). *Affective neuroscience: The foundation of human & animal emotion*. New York: Oxford University Press.

Panksepp, J., & Panksepp, J. (2000). The seven sins of evolutionary psychology. *Evolution and Cognition, 6*(2), 108–131.

Pellionisz, A., & Llinas, R. (1985). Tensor network theory of the metaorganization of functional geometries in the central nervous system. *Neuroscience, 7*, 2249–2970.

Perrett, D. I., & Rolls, E. T. (1983). Neural mechanisms underlying the visual analysis of faces. In J. -P. Ewert, R. R. Capranica, & D. J. Ingles (Eds.), *Advances in vertebrate neuroethology* (pp. 543–566). New York: Plenum Press.

Pfaff, D. W. (1999). *Drive: Neurobiological and molecular mechanism of sexual motivation*. Cambridge, MA: MIT Press.

Pfeiffer, J. (1982). *The creative explosion*. New York: Harper & Row.

Pinker, S. (1995). Beyond folk psychology. *Nature, 373*, 205.

Pinker, S. (1997). *How the mind works*. New York: W.W. Norton and Co.

Pinker, S. (2007). *The stuff of thought: Language as a window to human nature*. New York: Viking.

Place, U. T. (1956). Is consciousness a brain process? *British Journal of Psychology, 47*, 44–50.

Popper, K., & Eccles, J. (1977). *The self and its brain*. Berlin: Springer-International.

Posner, M. (1985). Chronometric measures of *g*. *Behavioral and Brain Sciences, 8*, 237–238.

Potts, R. (1988). *Early hominid activities at Olduvai Gorge*. New York: Aldine de Gruyter.

Prado, C. G. (1981). Sociobiology and materialist Theories of mind. *Dialogue, 20*(2), 247–268.

Pusey, A. (1980). Inbreeding avoidance in chimpanzees. *Animal Behaviour, 28*, 543–552.

Putnam, H. (1983). *Realism and reason*. Cambridge: Cambridge University Press.

Puts, D. (2010). The psychologic gambit declined—A review of 'endocrinology of social relationships'. *Evolution and Human Behavior, 31*(4), 306–308.

Puts, D., Gaulin, S., Sporter, R., & McBurney, D. (2004). Sex hormones and finger length: What does 2D:4D indicate? *Evolution and Human Behavior, 25*(3), 182–199.

Quine, W. V. O. (1960). *Word and object*. Cambridge, MA: MIT Press.

Quine, W. V. O. (1985). States of mind. *Journal of Philosophy, 82*(1), 5–8.

Quine, W. V. O. (1987). *Quiddities*. Cambridge, MA: Harvard University Press.

Quine, W. V. O. (1989). Mind, brain, and behavior. In A. J. Brownstein (Ed.), *Progress in behavioral studies* (Vol. 1, pp. 1–6). Mahwah, NJ: Lawrence Erlbaum Associates.

Quine, W. V. O. (1990). *The pursuit of truth*. Cambridge, MA: Harvard University Press.

Quine, W. V. O. (2008). *Confessions of a confirmed extensionalist and other essays*. Cambridge, MA: Harvard University Press.

Quine, W.V.O., & Ullian, J.S. (1978). *The web of belief* (2nd ed.). New York: McGraw-Hill.

Rantala, M., Eriksson, C. J., Vainikka, A., & Kortet, R. (2006). Male steroid hormones and female preference for male body odor. *Evolution and Human Behavior, 27*(4), 259–269.

Rantala, M., Eriksson, C. J., Vainikka, A., & Kortet, R. (2006). Male steroid hormones and female preference for male body odor. *Evolution and Human Behavior, 27*(4), 259–269.

Ridley, Mark. (1985). *The problems of evolution*. New York: Oxford University Press.

Ridley, Matt. (1996). *The origins of virtue*. New York: Penguin Books.

Rilling, J. K., Barks, S. K., Parr, L. A., Preuss, T. M., Faber, T. L., Pagnoni, G., et al. (2007). A comparison of resting-state brain activity in humans and chimpanzees. *Proceedings of the National Academy of Sciences, 104*(43), 17146–17151.

Robinson, S., & Manning, J. (2000). The ratio of 2nd to 4th digit length and male homosexuality. *Evolution and Human Behavior, 21*(5), 333–345.

Rolls, E. T. (1999). *The brain and emotion*. New York: Oxford University Press.

Rorty, R. (1965). Mind–body identity, privacy, and categories. *Review of Metaphysics, 19*, 24–54.

Routtenberg, A. (1978). The reward system of the brain. *Scientific American, 239*, 154–164.

Ryle, G. (1949). *The concept of mind*. Cambridge: Cambridge University Press.

Saussure, F. (1916/1998). *Course in general linguistics*. La Salle, IL: Open Court Press.

Savage-Rumbaugh, S., & Lewin, R. (1994). *Kanzi: The ape at the brink of the human mind*. New York: Wiley/Doubleday.

Searle, J. (1980a). Rules and causation. *Behavioral and Brain Sciences, 3*(1), 37–38.

Searle, J. (1980b). Minds, brains, and programs. *Behavioral and Brain Sciences, 3*(3), 417–457.

Searle, J. (1984). *Minds, brains and science*. Cambridge, MA: Harvard University Press.

Sellars, W. (1963). *Science, perception and reality*. London: Routledge and Kegan Paul.

Shoemaker, S. (2007). *Physical realization*. New York: Oxford University Press.

Simon, H. (1962). The architecture of complexity. *Proceedings of the American Philosophical Society, 106*(6), 467–482.

Skinner, B.F. (1948/1984). The operational analysis of psychological terms. *Behavioral and Brain Sciences*, 7(4), 547582.

Skinner, B.F. (1953). *Science and human behavior*. New York: MacMillan.

Skinner, B.F. (1957). *Verbal behavior*. New York: Appelton-Century-Crofts.

Skinner, B. F. (1977). Why I am not a cognitive psychologist. *Behaviorism, 5*, 1–11.

Skinner, B. F. (1979). *The shaping of a behaviorist*. New York: New York University Press.

Smart, J. J. C. (1959). Sensations and brain processes. *Philosophical Review, 68*(2), 141–156.

Sober, E. (1985). Methodological behaviorism, evolution, and game theory. In J. Fetzer (Ed.), *Sociobiology and epistemology*. Boston: D. Reidel.

Sober, E., & Wilson, D. S. (1998). *Unto others: The evolution and psychology of unselfish behavior*. Cambridge, MA: Harvard University Press.

Sperber, D. (1994). The modularity of thought and the epidemiology of representations. In L. A. Hirschfeld & S. A. Gelman (Eds.), *Mapping the mind: domain specificity in cognition and culture* (pp. 85–116). Cambridge: Cambridge University Press.

Sperber, D. & Girotto, V. (2003). Does the selection task detect cheater-detection? In K. Sterelny & J. Fitness (Eds.), *From mating to mentality: evaluating evolutionary psychology* (pp. 197–225). New York: Taylor & Francis Books Inc.

Sperber, D. & Wilson, D. (1986/1995). *Relevance: communication and cognition*. Cambridge, MA: Harvard University Press.

Sperber, D., Cara, F., & Girotto, V. (1995). Relevance theory explains the selection task. *Cognition, 54*, 3–39.

Sperry, R. W. (1974). Lateral specialization in the surgically separated hemispheres. In F. O. Schmitt & F. G. Worden (Eds.), *The neurosciences third study program* (pp. 5–20). Cambridge, MA: MIT Press.

Staten, H. (1986). *Wittgenstein and Derrida*. Lincoln: University of Nebraska Press.

Stein, L. (1978). Brain endorphins: Two mediators of pleasure and reward. *Neurosciences Research Program Bulletin, 16*, 556–567.

Steklis, H. D. & Lane, R. D. (2012). The unique human capacity for emotional awareness: Psychological, neuroanatomical, comparative and evolutionary perspectives. In S. Watanabe (Ed.), *Comparative perspectives on animal and human emotions*. New York: Springer (in press).

Steklis, H. D., & Kling, A. (1985). Neurobiology of affiliative behavior in non-human primates. In M. Reite & T. Field (Eds.), *The psychobiology of attachment and separation* (pp. 91–134). New York: Academic Press.

Steklis, H. D., & Walter, A. (1991). Culture, biology, and human behavior: A mechanistic approach. *Human Nature, 2*(2), 137–169.

Sterelny, K. (2003). *Thought in a hostile world: The evolution of human cognition*. Malden, MA: Blackwell.

Stich, S. (1978). Beliefs and subdoxastic states. *Philosophy of Science, 45*, 499–518.

Stich, S. (1983). *From folk psychology to cognitive science*. Cambridge, MA: MIT Press.

Stich, S. (1996). *Deconstructing the mind*. New York: Oxford University Press.

Suppe, F. (Ed.). (1977). *The structure of scientific theories*. Urbana: University of Illinois Press.

Symons, D. (1979). *The evolution of human sexuality*. New York: Oxford University Press.

Symons, D. (1992). On the use and misuse of darwinism in the study of human behavior. In J. Barkow, L. Cosmides & J. Tooby (Eds.), *The adapted mind* (pp. 137–159). New York: Oxford University Press.

Thornhill, R., & Gangestad, S. (2008). *The evolutionary biology of human female sexuality*. New York: Oxford University Press.

Thurstone, L. L. (1938). *Primary mental abilities*. Chicago: University of Chicago Press.

Tiger, L. (1994). A second look at biogrammar. *Social Science Information, 33*(4), 579–593.

Tiger, L., & Fox, R. (1971). *The imperial animal*. New York: Holt, Rinehart & Winston.

Tomasello, M. (2008). *Origins of human communication*. Cambridge: MIT Press.

Tooby, J., & Cosmides, L. (1989). Evolutionary psychology and the generation of culture, part 1: Theoretical considerations. *Ethology & Sociobiology, 10*, 51–98.

Tooby, J., & Cosmides, L. (1990). The past explains the present. emotional adaptations and the structure of ancestral environments. *Ethology and Sociobiology, 11*, 375–424.

Tooby, J., & Cosmides, L. (2005). Conceptual foundations of evolutionary psychology. In D. M. Buss (Ed.), *The handbook of evolutionary psychology* (pp. 5–67). Hoboken, NJ: John Wiley & Sons.

Toulmin, S. (1970). Reasons and causes. In R. Borger & F. Cioffi (Eds.), *Explanation in the behaviorial sciences* (pp. 1–25). Cambridge: Cambridge University Press.

Uttal, W. (2004). *Dualism: The original sin of cognitivism*. Mahwah, NJ: L. Erlbaum Associates.

Uttal, W. (2005). *Neural theories of mind: Why the mind–brain problem may never be solved*. Mahwah, NJ: L. Erlbaum Associates.

Vayda, A. P. (1995). Failures of explanation in Darwinian ecological anthropology: Part 1. *Philosophy of the Social Sciences, 25*(2), 219–245.

Vayda, A.P. (2008). Causal explanation as a research goal: A pragmatic view. In B. Walters et al. (Eds.), *Against the grain: The Vayda tradition in human ecology and ecological anthropology* (pp. 317–367). New York: Altamira.

Von Frisch, K. (1967). *The dance language and orientation of bees*. Cambridge, MA: Harvard University Press.

Walter, A. (1989). Pop goes the weasel. *Behavioral and Brain Sciences, 12*, 185–187.

Walter, A. (1997). The evolutionary psychology of mate selection in Morocco: A multivariate analysis. *Human Nature, 8*(2), 113–137.

Walter, A. (2000). From Westermarck's effect to Fox's law: Paradox and principle in the relationship between incest taboos and exogamy. *Social Science Information, 39*(3), 467–488.

Walter, A. (2002). Stone age detective's paradise: Investigating the mysteries of mind & culture. *Reviews in Anthropology, 31*(1), 29–61.

Walter, A. (2006). The anti-naturalistic fallacy: Evolutionary moral psychology and the insistence of brute facts. *Evolutionary Psychology, 4*, 33–48.

Walter, A. (2007). The trouble with memes: Deconstructing Dawkins' Monster. *Social Science Information, 46*(4), 691–709.

Walter, A., & Buyske, S. (2003). The Westermarck effect and early childhood cosocialization: Sex differences in inbreeding avoidance. *British Journal of Developmental Psychology, 21*, 353–365.

Wettstein, H. (1988). Cognitive significance without cognitive content. *Mind, 97*, 1–28.

Wettstein, H. (2004). *The magic prism.* New York: Oxford University Press.

Williams, G. C. (1966). *Adaptation and natural selection.* Princeton: Princeton University Press.

Wilson, E. O. (1975). *Sociobiology: The new synthesis.* Cambridge, MA: Harvard/Belknap Press.

Wise, R. A. (1978). Catecholamine theories of reward: A critical review. *Brain Research, 152*, 215–247.

Wise, R. A., & Bozarth, M. A. (1982). Action of drugs of abuse on brain reward systems: An update with specific attention to opiates. *Pharmacology, Biochemistry and Behavior, 17*, 239–243.

Wise, R. A., & Bozarth, M. A. (1984). Brain reward circuitry: Four circuit elements "wired" in apparent series. *Brain Research Bulletin, 12*, 203–208.

Wittgenstein, L. (1953). *Philosophical investigations.* New York: MacMillan.

Wolpoff, M. H. (1999). *Paleoanthropology* (2nd ed.). New York: McGraw-Hill.

Wolpoff, M. H., & Caspari, R. (1997). *Race and human evolution.* New York: Simon & Schuster.

Zajonc, R. (1980). Feeling and thinking: Preferences need no inferences. *American Psychologist, 35*, 151–175.

Index

A
Ache, 67
Aiello, L, 57, 65–67
Allen, M, 51
Alloparenting, 66–67
Anatomically modem humans, 58–59
Armstrong E, 30–31
Australopithecines, 66–67

B
Baker R, 40
Barash D, 7
Behaviorism, 26–27, 37–38, 70
Bellis M, 40
Bickerton D, 16
Binford L, 57
Block N, 20, 68
Blumenschine R, 57
Boden M, 59
Bonobos, 55
Bozarth M, 37
Buckner R, 31
Buller D, 33–34, 62–64
Buss D, 9, 30, 46
Byrne R, 54, 71

C
Carey S, 59
Cartesian dualism, 52, 68
Causey R, 67
Chalmers D, 19–20
Cheater-detection' module, 51, 61–62

Chimpanzees
Chimpanzees, 44, 54–56, 61
Chomsky N, 8, 12–13, 15–17, 42
Churchland P, 3, 5–6, 13–14, 18–20, 26, 28, 30, 37, 43–44, 69–71
Clark A, 18, 20
Communication, 16–17
Cosmides L, 9, 33–34, 36, 45–46, 51–52, 55, 60–62

D
Damasio A, 37, 51, 63–64
Davidson D, 19, 22, 24, 70
Dawkins, 7–8, 39, 65
de Saussure F, 6
Deacon T, 16, 42
Demarest W, 43
Dennett D, 2–3, 6–7, 10–11, 18, 26–27, 44
Derrida J, 6, 27
Design stance, 2, 44
Dixson A, 48
Domain-specific, 51–62
Drive reduction, 38–39
Dunbar R, 57

E
Ekman p, 33
Ellison p, 49
Emotion, 32–37
Evo-devo, 45
Evolutionary developmental systems, see evo-devo, 45, 46
Evolutionary stable strategy, 39, 41

A. Walter, *Evolutionary Psychology and the Propositional-attitudes*, SpringerBriefs in Philosophy, DOI: 10.1007/978-94-007-2969-8, © The Author(s) 2012

F
FA, see fluctuating asymmetry, 47, 48, 76
Falk d, 57
Feyerabend p, 18, 20, 70
Finger length ratio , 45–47
Fitch w, 17, 56
Flinn m, 75
Fluctuating asymmetry, 44, 47–49
Fodor j, 1, 6–7, 12, 17, 29, 33, 46, 49, 51, 58, 61–68
Folk psychology, 1–2, 18–19, 22, 24, 69–71
Fox r, 8
Frank r, 36
Free-will, 12

G
Gangestad s, 44, 47–49
Gardner h, 59–60
Garver-apgar c, 48
Gazzaniga m, 51, 54, 64
Gelman s, 26, 71
Gentner d, 65
Girotto v, 61–62
Globus h, 28
Gould j, 4
Gould s.j, 65–67
Gray r, 45, 49
Greenfield p, 54
Griffiths p, 33, 36, 45
Grossberg s, 5, 43

H
Hadza, 67
Hauser m, 17, 56
Heatherton t, 31
Hellige j, 51
Homo habilis, 57
Homo sapiens sapiens, 31, 57–58
Hughes, 60–61
Hume, d, 6

I
Imperatives, 28–29
Inbreeding avoidance, 43–44, 52
Indicatives, 4–5, 28–29
Information processing, 9, 12, 28, 42
Information processing models, 2, 28, 42
Intensionality, 27
Intensions, 7, 22
Intentional stance, 2, 5

Intention, see intentionality, 3, 11, 24
Intentional systems, 7, 10–12
Intentionality, 2–7, 20–23, 27–29, 42–44, 49
IPM, 12, 13, 42
Isaac g, 67
Izard c, 33

K
Kalke w, 67
Karmiloff-smith a, 34, 54, 61
Kim j, 18, 26, 70
Kitcher p, 43, 68
Kluge, 34
Kuhn t, 69

L
Lakatos i, 69
Lakoff g, 65
Language of thought, 1, 6, 12, 22, 27–28, 32, 49, 67
Learning rule, 14
Ledoux j, 32, 35–37
Leslie a, 31–32, 71
Lewontin r, 65–67
Lickliter r, 46
Limbic cortex, 34–35
Llinas r, 15
Lycan w, 70

M
Maccorquodale k, 12
Machiavellian social intelligence, 55, 71
Manning, 45–47
Marcus g, 33–34, 54, 56, 64
Maynard-smith j, 40, 60
Mccarthy j, 20
Mclaughlin b, 19–20
Mechanist stance, 27, 44
Mentalese, 13, 53
Meta-representation, 58, 64
Miller g, 9
Millikan r, 3–4, 6, 24, 29, 43
Minsky m, 18
Mithen, 54–65
Modularity, 9, 33, 51–61, 63–65
Morillo c, 37
Motivational endowments, 28–31, 44
Multiple intelligence theory, 59–61, 63–65

N
Natural selection, 2, 42
Neanderthals, 57–58

O
Oldowan, 56
Olds j, 37
Operant conditioning, 11, 24

P
Panksepp j, 30, 32, 38
Pellionisz a, 15
Pfaff d, 30, 38–39
Pfeiffer j, 59
Phenomenological fallacy, 21
Physical stance, 2, 5, 28
Pinker s, 14, 18, 52–53, 65
Place u.t, 20–21
Platonic, 6, 18, 27, 46, 51
Platonism, see platonic, 6
Plomin r, 60–61
Posner m, 60
Potts r, 57
Prado c, 24
Propositional-attitude, 1, 3, 10, 27, 29, 36–37,
        40, 44–45, 49, 67
Proximate mechanism, 2, 5, 7–10, 23, 25, 28,
        31, 36, 40, 42–44
Psychologic gambit, 44–45, 49
Pusey a, 44

Q
Qualia, 18, 21
Quine, w.v.o, 6–7, 17, 19, 21, 26–27,
        44, 70, 78

R
Rantala m, 48
Reward event theory, 37
Ridley m, 39
Robinson s, 47
Rolls e, 30, 39–41
Rorty r, 20
Routtenberg,a, 37
Rules and representations, 12, 14–15, 17, 41–42
Ryle g, 6

S
Savage-rumbaugh s, 55
Searle j, 3, 15, 20–23, 44, 49

Shoemaker s, 18, 20
Skinner b.f, 11–12, 23, 26, 38
Smart j.j.c, 6, 18, 20–21
Sober e, 10, 43
Social intelligence, 54, 62
Sociobiology, 3, 7–8, 42, 67, V, VI
Spelke s, 59
Sperber d, 59, 61–62, 64
St, thomas aquinas, 5–7
Stein l, 38
Steklis h.d, 31, 63
Sterelny k, 4, 8, 12, 14, 29, 36, 43, 49
Stich s, 2, 19, 70
Suppe f, 70
Symons d, 25
Szathmary e, 60

T
Theory of mind, 31–32, 60–61, 71
Thornhill r, 45–48
Thurstone l, 60
Tiger l, 8
Tomasello m, 16–17, 42, 56
Tooby j, 9, 33–34, 36, 45–46, 51, 60–62
Toulmin s, 22

U
Ultimate causation, 2, 5, 7
Uttal w, 27, 38

V
Vayda a.p, 9–10, 43, 67

W
Waggle dance, 4
Wason task, 52, 61–62
Wettstein h, 16
Wheeler p, 65
Whiten a, 54, 71
Wilson e.o, 7
Wilson d, 62
Wilson d.s, 10
Wise, 37–38
Wittgenstein l., 6, 16, 27, 71
Wolpoff m, 57

Z
Zajonc r, 32–33